BIOLOGICAL WEAPONS

BIOLOGICAL WEAPONS

From the Invention of State-Sponsored Programs to Contemporary Bioterrorism

JEANNE GUILLEMIN

COLUMBIA UNIVERSITY PRESS NEW YORK

COLUMBIA UNIVERSITY PRESS
Publishers Since 1893
NEW YORK CHICHESTER, WEST SUSSEX

Library of Congress Cataloging-in-Publication Data
Guillemin, Jeanne, 1943–
Biological weapons : from the invention of state-sponsored programs to
comtemporary bioterrorism / Jeanne Guillemin.
p. cm.
Includes bibliographical references and index.
ISBN 0–231–12942–4 (cloth : alk. paper)—ISBN 0–231–50917–0 (elec.)
1. Biological warfare—History. 2. Bioterrorism. I. Title.
UG447.8.G85 2004
358′.3882′09—dc22 2004051911

∞

Columbia University Press books are printed on permanent and
durable acid-free paper.
Printed in the United States of America

c 10 9 8 7 6 5 4 3 2 1

CONTENTS

PREFACE

The aim of this book is to bring historical context to present concerns about biological weapons and the potential for bioterrorism. The history of state-sponsored biological weapons programs is much deeper than most people are aware. During the twentieth century, major state powers (France, Japan, the United Kingdom, the United States, and the Soviet Union) developed biological warfare programs in such great secrecy that many documents remain unavailable. By 1972, with the signing of the Biological Weapons Convention, the state programs were ended, or from that point they became cloaked in even greater secrecy. With a single exception, the state programs never led to the use of biological weapons in war. From 1939 to 1945, the Japanese Imperial Army surreptitiously spread plague and cholera in China, under the guise of natural outbreaks. Even so, the world has never witnessed any battlefield exchanges or aerial bombings with germ weapons. After 1945, nuclear weapons overshadowed the threat of biological weapons, until the Cold War ended. Then the untested potential of biological weapons emerged as a new order of threat, appearing more technically accessible than either nuclear or chemical weapons.

The lack of use of biological weapons in war was not for want of great effort and enormous funding invested in state programs. Advocates for using biology to create a new class of weapons initially envisioned delivery systems for pathogenic aerosols that mimicked those for chemical weapons, which were mainly bombs that generated aerosols intended to kill or disable troops in a local area. This vision was quickly replaced by the concept of creating huge clouds of germs that would drift with the wind and infect people over areas of thousands of square miles. The scientists and civil and military leaders who believed in the future of biological weapons saw their potential for fulfilling the goals of total war, that is, for the mass killing or debilitation of enemy civilians. It is often asked why biological weapons are different from any other means of destruction. The answer is that they are the only ones devised expressly to kill defenseless humans and animal and plant life, with little real battlefield potential in modern war.

We live in the aftermath of the cumulative technical achievements of now-defunct state programs, especially those of the United States and the Soviet Union. Total-war doctrine is an anachronism in advanced industrial nations, but threats of total violence against civilians persist around the world. With the end of the Cold War, the framework for analyzing conflict shifted to unrestrained regional conflicts and religious or nationalist extremism at times allied with global terrorism. The fear has grown that small states and terrorists will gain possession of the technologies of total war and use them. In the 1980s, in South Africa and Iraq, we saw the transfer of the technology of biological weapons to highly contentious world areas where, had those programs been successfully developed, racial and ethnic animosities would likely have precipitated large-scale use. In the 1990s, with the Aum Shinrikyo cult, we saw an apocalyptic vision of destruction turn to experiments with biological and chemical weapons, culminating in the sarin nerve gas attacks in the Tokyo subway in 1995.

This book seeks first to explain the different historical circumstances in which scientists and political leaders were able to mobilize the resources for biological weapons development, after the use of these weapons had been banned by the 1925 Geneva Protocol. These circumstances include the very different positions of France and Japan in the years after World War I, the recruitment of British and American scientists during World War II, and then the Cold War dynamics that propelled the American and Soviet biological warfare programs. They also include the aforementioned examples of Iraq and South Africa, whose ambitions to acquire biological weapons (along with nuclear and chemical ones) were based more on regional than Cold War politics.

What made biological weapons a failed military innovation? Even when pathogens were prepared, munitions were tested, and attack plans were formulated, political and military authorities did not cross the line and use them for strategic wartime attacks, when they very well might have. The history of chemical weapons can shed some light on why. In World War II, Allied and Axis military commanders resisted using chemical weapons and thereby left an entire class of armaments, tested in battle during World War I, on the shelf. A combination of factors influenced their choice. Legal restraints, the force of public opinion, and technical drawbacks (including the enemy's defenses, using masks and protective garb) each worked against the reintroduction of chemical weapons to the battlefield.[1]

The prospects of retaliation and escalation also kept chemical weapons unused, and this same reckoning of consequences has likely prevented the use of nuclear weapons. In all wars, the advantage of surprise attack can be weighed against expectations of how an enemy might respond. The process of second-guessing who would dare strike first leads to calculations of the contingencies of retaliation and counterretaliation that, in turn, raise specters of out-of-control escalation and self-defeating conflagrations. It could be that such weapons are not worth having at all.[2] When biological weapons were developed for possible retaliation against an enemy thought to be similarly armed, they fit this model of restraint.

Nevertheless, germ weapons were developed and laws and political circumstances offered no guarantees against their use. Historians may seek neat explanations, but the element of uncertainty was always there. The 1925 Geneva Protocol and the 1972 Biological Weapons Convention expressed norms that have had widespread support, including in the military, but did not prevent arms buildups. Public opinion against biological weapons counted effectively during the 1960s in the United Kingdom and the United States, but often the public was ignorant of the programs or incapable of influencing policy. As for technological shortfalls, test failures were frequently discounted by military promises that the weapons' vast potential could be achieved, given more time and money. Political and military authorities differed unpredictably when it came to calculating the consequences of using biological weapons. Some saw them as potentially a great advantage, precisely because they were unusual; others, like Hitler, declared them anathema and refused their advocates' support. Today the problem is how, instead of leaving safeguards to the whims of fortune, more secure methods can be found to prevent the proliferation and use of biological weapons.

In its final chapters, this book describes the new subject of bioterrorism and traces at least the beginnings of the present era, in which domestic preparedness and homeland security are major policy issues, particularly in the United States. It has been nearly fifteen years since, with the Cold War over, American politicians envisioned bioterrorism as a major threat to national security and laid the groundwork for domestic preparedness programs. After the 9/11 terrorist attacks and the anthrax letters discovered a month later, the nation went on guard, with some cities, like Washington and New York, perpetually on high alert. In 2003, with the establishment of the Department of Homeland Security, the United States created a unique institutional base for civil defense. This organizational innovation and the larger homeland security effort could put America into permanent emergency mode, verging it toward the edgy "risk society" that advanced industrial nations become when social policies founder.[3] If the public is in danger from biological weapons, how can it best be protected, and how can it act to save itself? The answers are far from obvious.

One of the major homeland security directives is to use technology derived from medical research to protect civilians against bioterrorism. The use of biology to defend civilians against any and all biological agents is a daunting project that imposes new security restrictions more familiar to physicists in the defense establishment than to biologists. Simultaneously, the biodefense initiative has provoked new awareness of the international dimensions of modern biology, a knowledge enterprise that no nation, not even the powerful United States, can control for itself. The global transfer of new biotechnologies is inevitable, and with it will come greater mastery of human physiology, including cognition, development, and reproduction. The question is not just how Americans might best defend themselves against bioterrorism, which needs serious consideration. The bigger problem is how we are going to prevent the misappropriation of these innovations wherever it might happen, no matter who might be the targeted victims.

The study of biological weapons combines knowledge from disparate fields: biology, medicine, military history, politics, law, and ethics. This book intends to give the reader a basic literacy in this complex area, with the sources cited in footnotes as guides to further reading—which is highly recommended. The entire subject of biological weapons is characterized by an unusual degree of misinformation and even disinformation. Almost every fact in this book has a hundred-page backstory, with greater nuance and depth than a brief overview can provide. As scholars continue their work, more information and analysis will likely turn today's accepted wisdom on its head. This progress will be a healthy sign for the field, whose subject matter has often been exploited for the frightening and sometimes entertaining effect it has on the imagination.

A number of important institutes and programs have taken responsibility for educating the public about biological weapons. The resources of the Harvard–Sussex Program, the Monterey Institute, the University of Bradford (UK) Department of Peace Studies, the Federation of American Scientists, the Stockholm International Peace Research Institute, and others can be located electronically. The US government and its agencies are important resources for information on biological weapons history and on American programs to counter bioterrorism. The UK Public Record Office (PRO) in

London (now the National Archives) is a treasure trove of formerly classified documents on the British and other programs.

Before the world was conscious of any threat from bioterrorism, a select handful of people committed themselves to restraining biological and chemical weapons proliferation and use. Their written contributions, much in evidence in this book, and their activism set a high standard for any efforts that follow. My own background is in research on emergency medical systems. I came to the study of biological weapons primarily as an anthropological field investigator in controversies regarding treaty violations. Beginning in 1982, I joined an independent academic inquiry into allegations of Soviet complicity in using mycotoxins in Southeast Asia. Later, I was involved in a similar inquiry into the 1979 anthrax outbreak in the former Soviet city of Sverdlovsk. My present work is on the 2001 anthrax postal attacks, an important impetus for US defensive policies against the biological weapons threat.

Brian Balmer, Steve Black, Paul Doty, Martin Furmanski, Ben Garrett, Gregory Koblentz, Olivier Lepick, Richard Samuels, Nicholas Simms, and Nikita Smidovich read and commented on selected chapters. Dena Briscoe and Terrell Worrell of Brentwood Exposed read and corrected portions of the book relating to the 2001 anthrax postal attacks. Julian Perry Robinson and Matthew Meselson, co-directors of the Harvard–Sussex Program, read the book manuscript in earlier drafts and made many constructive suggestions. I much appreciated all this help. New or persistent errors or misconceptions are, of course, my responsibility alone.

Many people have educated me with valuable insights based on professional experience, often with points of view different from my own. Among them are Elisa Harris, Jessica Stern, Joshua Lederberg, Spurgeon Keeney, Henry Kissinger, Sheila Jasanoff, Juliette Kayyem, Graham Allison, Benjamin Garrett, Gerald Holton, James LeDuc, Peter Brown, Catherine Kelleher, Irving Louis Horowitz, Han Swyter, Paul Schulte, Donald Mahley, Robert Mikulak, Erhard Geissler, Yi-fu Tuan, Rajan Gupta, Paul Farmer, David Scott, Ken Alibek, Edgar Larson, Joseph Jemski, and Michael Bray. My colleagues at the Security Studies Program at MIT and at the Dibner Institute for the History of Science and Technology provided an exceptional environment for completing this project, which was conceived in the aftermath of 9/11 and the anthrax postal attacks.

The staff of the Harvard–Sussex Program at the Science Policy Research Unit, Sussex University, generously aided this research. I thank HSP's Sandy Ropper at Harvard for her invaluable archival research and Barbara Ring for her long-time organizational assistance. I also thank Jessica Brennan, Autumn Green, Toni Vicari, and Austin Long for research and editing help.

Funding for my research and writing on contemporary events was provided by the John D. and Catherine T. MacArthur Foundation, the Program on Global Security and Sustainability. In 2002, the Dibner Institute funded my archival research. At Boston College, dean of Arts and Sciences Joseph Quinn, dean of the Graduate School Michael Smyer, and Sociology chairman Stephen Pfohl supported my pursuit of this leave-time project. James Warren, my editor at Columbia University Press, has been a perceptive, knowledgeable guide. My literary agents, Jill Neerim and Ike Williams, gave the best of professional advice and service.

This account of the long years of state sponsorship of biological warfare programs makes for disconcerting reading. Yet it is worth remembering that throughout history much more human energy has been devoted to fighting major epidemics than to deliberately causing them. The grim history of biological weapons should remind us that over the years the world's judgment of these programs radically changed for the better, as states that once legitimized them finally rejected them as morally repugnant.

The secrecy that surrounded biological weapons programs posed a special danger to people as they went about their ordinary lives. In the past, government authorities were erratic in setting their priorities, covertly expanding or contracting their plans to start deadly epidemics, targeting civilians without hesitancy or reflection. Secrecy can increase risks even when programs are purely defensive, for the negative effect it has on communication between government officials and the public. In any disease outbreak, one's instinct is to have all possible accurate, relevant information, not selected data filtered through a national security expert. This instinct should be encouraged. Disease is a most personal experience and, to make the right self-protective decisions, the individual must be fully informed of the contingencies of disease transmission in advance, not when sirens are blaring.

As with other epidemics, deliberately inflicted disease can be protected against and cured. But technology is never the complete answer; to an important degree, civilians will remain vulnerable to a determined biological attack. For this reason we need political restraints that lower risks. Biological weapons can be controlled by strong international treaties and law enforcement that engage states in protecting civilians and societies, with the aim of doing so universally, not merely selectively by nationality, and with openness as a principle goal. Any policies to reduce the threat of these weapons should be based on public understanding of how political choices can increase the risks of intentional epidemics and that, conversely, the political means for reducing those risks are available. It is in the spirit of furthering this understanding that this book is written.

BIOLOGICAL WEAPONS

INTRODUCTION

The history of biological weapons programs is a repetitive spectacle of biological science put to its worst use, of threats imagined and ignored, and of government secrecy increasing annihilating risks to civilians. Along the way, legal and technical restraints, civic awareness, and the decisions of key political actors have kept this innovative class of weapons from the destructive strategic uses its advocates envisioned. This book is about the twentieth-century incorporation of biological weapons into the arsenals of industrial states and its implications for present times, when new technologies and persistent political animosities may allow even more ominous threats than in the past.

The rise of state-sponsored programs is inseparable from the great conflicts among advanced industrial nations and the general search for technological advantage in warfare. France in the 1920s, followed by Japan in 1934 and subsequently by the United Kingdom, the United States, and the Soviet Union, had serious expectations that biological weapons could win wars. Each invested heavily in research and development to explore this possibility. The prior development of chemical weapons during World War I aided technological advances in biological weapons; the history of these two classes of weapons is often overlapping. The introduction of aerial warfare and long-range bombers expanded the vision of how to devastate enemy civilians with disease. By the end of World War II, the United States was well on its way to industrial-scale production of munitions. It then pursued the goal of making biological attacks as destructive in scope as nuclear bombs. The Soviet Union followed with even greater strategic capacity.

By the end of the twentieth century, these programs were abandoned, with the Soviet Union as the last holdout. In their aftermath, though, the threat has devolved to other states that might also see advantages in biological weapons to settle their conflicts or to terrorists that might acquire them to perpetrate attacks on major powers, especially the United States. Whether possessed by states or by terrorists, biological weapons are a particular menace to civilians. From the beginning, advocates of biological weapons programs envisioned the mass killing of noncombatants as the means to victory. No holds were barred. The destruction of livestock and crops essential to civilians was also part of strategic attack plans. Biological weapons continue to lend themselves to indiscriminate attacks on defenseless populations. Compared to nuclear weapons, the skills and technologies to inflict intentional epidemics are readily available. Hatreds that dehumanize entire national populations or ethnic or religious groups are as prevalent as ever.

In addition to political discord, the future of biological weapons is bound to be influenced by advances in biotechnology that will expand the dimensions for destructive use, even as they increase the options for medical interventions. Should the deliberate infliction of disease and other hostile applications of biotechnology come into widespread use, the nature of war and the future course of civilization itself could be disastrously altered.

This introduction presents a brief synopsis of the early context for political restraints against biological weapons, when they were barely distinguishable from chemical weapons, and before states began to invest in them. An outline of the contents and themes of chapters then follows. Before we begin this review of how civilized nations have verged on the use of biological weapons, a few basic terms need definition:

1. *Biological warfare* refers to the military employment of microbial or other biological agents (including bacteria, viruses, and fungi) or toxins to produce death, temporary incapacitation, or permanent harm in humans or to kill or damage animals or plants for a military objective. Toxins are harmful substances obtained from microbial or other living organisms or their chemical analogues, whatever their origin or method of production.

2. *Biological weapons* refer to munitions, equipment or other means of delivery including bombs, aircraft spray tanks and other devices, intended for use in the dissemination of biological agents and toxins for hostile purposes. The principal means of dissemination are as an aerosol to be inhaled by a target population or as a spray to be deposited on crop plants. An aerosol is a suspension in air of particles so small that they travel with air currents instead of settling to the ground.

3. *Biological weapons agents* refer to microbial and other biological agents and toxins when intended for use in biological weapons. These are also referred to as select agents, with secondary reference to the diseases they cause; for example, *Bacillus anthracis* is the bacterium that causes anthrax.

Chemical and Biological "Poisons"

Poison and infectious disease were not clearly distinguished from each other until the establishment of the germ theory of disease at the end of the nineteenth century. In contrast to state-sponsored programs, preindustrial societies offer fascinating but only sporadic examples of the deliberate use of poison by armies. Two thousand years ago the Roman author Valerius Maximus summarized the preference: "War is waged with arms, not poison." Since then, the most common exceptions have been attempts to pollute enemy wells with animal carcasses or the pitching of the corpses of plague victims into fortressed cities, as in the 1346 siege of Caffa. In one notorious and more modern example, in 1763 the British army used a few infected blankets to start a smallpox epidemic in an enemy American Indian tribe.[1]

In the late nineteenth century, researchers studied disease-producing microorganisms and toxins (such as botulinum toxin, a highly poisonous substance produced by a bacterium) almost solely with the aim of preventing and curing infectious disease. In 1874, the exploitation of microbiology for military purposes was barely imagined when the states represented at the Brussels Conference on the Laws and Customs of War agreed to prohibit the use in war of "poison or poisoned weapons." Rather, awareness of the rapidly growing importance of chemistry and the chemical industry raised concerns about an entirely new kind of weapon: noxious clouds. In 1899, European delegates to the first international peace conference at The Hague agreed "to abstain from the use of projectiles, the object of which is the diffusion of asphyxiating or deleterious gases."[2] The Hague Convention of 1907, signed by France, Germany, Great Britain, and most other European powers, reaffirmed the prior bans.

These simple prohibitions, none of them backed by provisions for enforcement, were to no avail. In World War I, both sides made massive use of chemical weapons, including chlorine, phosgene, mustard gas (actually a liquid), tear gases, and various other toxic chemicals. In the first large-scale use of gas, at Ypres, Belgium, in 1915, chlorine released by German troops from thousands of cylinders was carried by wind across the battlefield to the

Allied positions. Lacking any protection and taken by surprise, the defenders, mostly French colonial troops, were routed. Apparently not expecting such success, the German forces failed to exploit the temporary breakthrough, and then the Allies quickly learned to respond in kind.

An important aspect of chemical weapons, one that diminished their battlefield value, was that simple individual defenses against their use could be invented. As the war progressed, fairly effective masks and clothing were devised and troops were taught how and when to use them. Although such protective equipment and training prevented gas from playing any decisive role in the war, chemical weapons nevertheless inflicted great suffering on the unprepared and, in many cases, long-lasting debilitation. After World War I, chemical weapons were generally regarded as among the great horrors of that conflict, worse for some than the other traumas of trench warfare. When British poet Wilfred Owen described terrified battlefield comrades "drowning in a yellow sea" of chlorine gas, veterans and their families understood the horror and so did politicians.

The 1922 Treaty of Washington banned the use in war of "asphyxiating, poisonous or other gases, and all analogous liquids, materials or devices." Although the US Senate gave its advice and consent to ratification, the 1922 treaty did not enter into force because France objected to its provisions on submarine warfare. Nevertheless, the chemical provisions of the Washington treaty became the template for the 1925 Geneva Protocol. This treaty, signed on June 17, 1925, entered into force on February 8, 1928, and now has 132 state parties, including all major nations of the world. The Geneva Protocol prohibited the use of chemical weapons and extended the prohibition to include "the use of bacteriological methods of warfare."

The United States signed the Geneva Protocol but failed to ratify it. During the 1920s, an isolationist Senate leadership, the American Legion, and the American chemical industry organized successful opposition to the protocol, and the Senate took no action to approve it. Instead, the United States professed to abide by the treaty's principles and over the years claimed the latitude to interpret what the Geneva Protocol meant.

Biological weapons were left unused in modern war, with two exceptions. Japan, which also did not ratify the Geneva Protocol, developed a biological program in Manchuria from 1934 to 1945 and its army there was unique in perpetrating the intentional spread of epidemics. Using crude methods, such as plague-infected fleas and contamination of food and water, the Japanese Imperial Army caused significant and recurrent illness among Chinese civilians and may have been successful in sabotage against Soviet troops following a battle at a northern border. The Japanese intended that the outbreaks they started would appear due to seasonal or other natural causes, and for years uncertainty surrounded these attacks.

The other exception, even more difficult to document, was the use of biological agents in special covert operations. The German attempt to infect pack animals with anthrax and glanders during World War I is perhaps the best recorded example of this kind of sabotage. More important, the advanced state programs in the West and the Soviet Union far exceeded the technical achievements of the Japanese by perfecting bombs and other aerial delivery systems and developing virulent agents. While all the advanced programs had clandestine potential, each put a much greater priority on strategic scale.

The Geneva Protocol banned the first use of biological weapons but was silent about the acquisition of this new class of weapons. It therefore allowed a state party to prepare itself for reprisal in kind if an enemy state attacked it with chemical or biological weapons. The concept of reprisal in kind was much debated. Ostensibly, it allowed limited or symmetrical use of a prohibited weapon, if it were undertaken to persuade an adversary using the weapon to stop. Whether reprisals in kind against civilian populations fit this exception was a matter of contention. If a nation's population were attacked, was it allowed then to seek retribution in kind on enemy civilians?[3] The line between defensive and offensive preparation could easily become vague. Since the scale of an enemy's capacity was likely to be unknown, there were incentives to maximize one's own capacity for surprise attack, provided the enemy's masks, suits, and medicines were ruled out as technical defenses that would defeat the plan.

Most of the major parties to the Geneva Protocol submitted formal reservations asserting that they would not be bound by the treaty if they were first attacked with chemical or biological weapons. For defensive purposes, therefore, they preserved the latitude for counterattack. Reservations were also added by individual states, starting with France, that the treaty would cease to bind them if they were attacked by armed forces allied with an enemy state, not just that state itself. Thus, the reservations allowed states to make broader claims for exempting themselves, in self-defense, from the treaty's ban on use.

For chemical weapons, the buildup of an arsenal might also deter an enemy from first strike. Some evidence exists that this is exactly what happened in World War II, when the British and Americans threatened serious reprisal should Germany use chemical weapons on any front. With biological weapons, no government ever overtly trumpeted its retaliatory capacity. The existence of biological weapons programs was generally more secret and their technical difficulties more acute than for chemical weapons. Suspicion played a strong part in justifying defensive programs that soon became offensive in capacity and then were secretly developed for first-strike advantage. The threat of an opponent's biological weapons was almost always invoked to accelerate the creation and expansion of the major programs.

The Allure of Bacteriological Weapons

The peace movement that followed World War I, which led to the League of Nations and the Geneva Disarmament Conference, was defeated by the competing vision that modern technologies should be fully exploited for national defense and expansion. The machine gun had changed battlefield tactics in Europe and the colonies. Other new technologies promised even greater efficiency. The advocates of air war, for example, argued that aerial bombing could resolve international conflicts more quickly and avoid the protracted battlefield slaughter of the Great War. The Italian military analyst Guilio Douhet and American Col. Billy Mitchell believed that heavy bombers and improved technology for aerial targeting could eliminate extensive land warfare; Douhet in particular saw a role for gas bombing as a way to shorten war with overwhelming surprise attacks on the enemy.[4] Even before it was technically possible, the idea of planes spreading invisible clouds of germs over enemy factories and cities stirred the imaginations of the military-minded scientists who conceived the first biological weapons programs in the years leading to World War II.

To their early advocates, chemical weapons and then bacteriological weapons, as they were called, were viewed as modern applications of advanced scientific knowledge that would cause mass casualties more efficiently than conventional arms, without tearing the enemy limb from limb or exposing the attacker to great harm. In theory, air warfare allowed pilot and crew to escape risk while troops on the ground and other remote targets absorbed the impact of attack.[5]

In the history of both chemical and biological weapons, their vaunted modernity was used by advocates to appropriate moral considerations. During World War I, the German government and press argued that chemical weapons were advantageous because they did not destroy buildings or bridges and were a humane alternative to high explosives because they avoided battlefield blood and gore. These arguments, repeated by chemical weapons advocates elsewhere, rationalized chemicals as "a higher form of killing."[6] Similarly, during World War II, some British experts saw biological weapons as a more humane way not of killing soldiers but of killing civilians already doomed, it was thought, by aerial attacks with high explosives.[7]

The special advantage associated with biological weapons over chemicals was their perceived potential for large-scale attack on enemy cities and industrial centers. Starting with the French in the 1920s, biological weapons were recognized as having little or no battlefield utility. Pathogenic aerosols were difficult to target and their impact was slow and unpredictable, given the speed and impetus demanded in ground warfare. They might lend a one-time surprise advantage, but, as with chemical weapons, troops could be well defended. If enemy soldiers were immunized or had good facemasks or other protection, a biological attack could be a waste of crucial wartime resources.

In contrast, the dissemination of disease agents by airplanes over enemy urban and industrial targets promised to destroy the civilian workforce and economic infrastructure of an enemy country. In this way, biological weapons were consonant with total war doctrine as it emerged in the twentieth century, when armed hostilities between industrial nations blurred the boundaries between enemy soldiers and civilians.[8] The offensive biological warfare programs were based on a dehumanization of enemy civilians as a mass that, for purposes of war or domination, had to be efficiently and predictably infected with disease. The impact behind the lines could be slower and less precise than at the battlefront, for it was an attack on an entire social structure and the masses of civilians that sustained it. The Allied forces' "carpet" bombings of German and Japanese cities in World War II were no different in principle from the Allied plans for biological warfare attacks. In parallel fashion, Imperial Japan's covert spread of diseases among Chinese civilians was consonant with its simultaneous wartime bombings of Chinese and other Asian cities.

Within biological warfare programs, so-called nonlethal agents were represented as both benign and efficient. During the 1950s, the United States, in addition to work on highly lethal biological weapons, created a large stockpile of bombs containing the pathogen of the disease brucellosis, an agent already well explored. Brucellosis causes fever, drenching sweats, and highly incapacitating headache and back pain, usually lasting for weeks or longer, and has an estimated fatality rate in natural outbreaks of about two percent. This estimate, though, meant that in a large city tens of thousands would die, so that the weapon was actually lethal in its effect. If the target population included high numbers of people already at risk (suffering from wartime deprivations, such as malnourishment, loss of housing, and poor medical care, and also the elderly, children, pregnant women and the sick), the fatality rate would be much higher than predicted.

For each offensive program that was planned, military defenses—protective masks, vaccines, and drugs—were included to reduce the impact of surprise attack. Was it possible, though, to systematically protect hundreds of thousands of civilians from enemy attack? Conventional civil defense measures, such as wearing gas masks or taking shelter, were as far as most national efforts went, and these would have been of limited value in the event of a biological attack.

When developing germ weapons in 1940, in anticipation of a German attack, the British took the position that a generally healthy population well served by hospitals was the best defense against all and any wartime attacks. After World War II, the UK Ministry of Health considered that given the necessary additional staff and centers, "the logistics of mass immunization were daunting," and, in addition, the authorities would have difficulty persuading the public to cooperate "without any imminent danger of war."[9] The goal of protecting an entire national population against potential germ warfare emerged most strongly in the 1990s when the United States, Israel, Japan, and other states began reacting to the perceived threat of bioterrorism. At this point, in the aftermath of the major state programs, the framework for anticipated attacks on civilians had radically altered. What stayed the same was the relatively slow impact of infectious disease, which allows public health planning and response.

Combined Restraints on Use

The decline in use of chemical weapons and the lack of use of biological weapons are unsolved puzzles in military history. Both sides in World War II stockpiled chemical weapons, yet the major powers refrained from using them. The restraints that prevented a reprise of chemical weapons attacks in that era are very likely the same that, in various combinations, subsequently prevented major states from using biological weapons.[10] International treaty and customary law played a part, making some nations hesitant to cross the line to illegal use even when there were no overt sanctions. Unless a state can afford to retreat from world opinion, the use of chemical weapons identifies a user state with a kind of ruthless barbarism, as it did when Iraq used them against defenseless Iranian troops, including children, in the 1980s Iran-Iraq War. Within states responsive to public opinion, the opinions and dissent of citizens can weigh heavily against unpopular weapons, which is what chemicals became after World War I. Chemical weapons became unpopular again in the 1960s, during the Vietnam War, with elected representatives, the press, and scientific advisors weighing in first against tear gas and herbicides and then against the entire class of biological and toxin weapons.

Top leadership in government made a difference in preventing the use of chemical weapons during World War II and in keeping biological weapons from being assimilated by the military. President Franklin Roosevelt strongly believed that both chemical and biological weapons were uncivilized and should never be used. Strangely enough, Adolf Hitler, who did not hesitate at mass murders by poison, was also averse to chemical and biological weapons. In 1969, President Richard Nixon renounced biological weapons on behalf of the United States and so broke the connection between the military and advancing American and Western biotechnology. In the 1990s, President Bill Clinton's vision of bioterrorism had, by the end of the century, laid the groundwork for domestic preparedness initiatives and increased intelligence surveillance to discourage attack.

Over the years, a key restraining role was played by the military, which generally shared an aversion to chemical and biological weapons. Military commanders, if not morally repulsed by them, at least preferred weapons that were predictable and had immediate effects. Admiral William Leahy, Roosevelt's and later President Harry Truman's top military advisor, was vocal in his opposition to weapons that target civilians. Others who accepted nuclear weapons were set against chemical and biological programs.

Chemical weapons, pound for pound less effective than biological agents, had similar disadvantages with regard to targeting and these restrained their use. Biological weapons had their own order of problems, in addition to being diffuse "noxious clouds." A major technical difficulty was determining the dose response for an agent, for example, how many inhaled spores of anthrax could be counted on to kill an exposed population. The dose response calculation was fundamental to assessing the worth of the entire biological weapons project. On it rested the construction of predictable weapons and, on that, the calculation of munitions requirements and aerial sorties over enemy targets. Calculation of dose response ultimately determined the entire allocation of resources and the formulation of military doctrine whereby the armed forces would be organized and trained to respond to threats and be able to discern opportunities to use biological weapons.[11] Animal research was repeatedly conducted to ascertain to what extent biological weapons agents might infect an enemy population. In only a few instances, for example, in research on tularemia and Q (query) fever, which could be reliably cured with antibiotics, the US military was able to engage in human-subjects research that better approximated attack scenarios. Otherwise, much guesswork went into estimating the effects of a biological aerosol.

Another major problem was achieving accurate or even approximate targeting. Once released, germ aerosols move with air currents and, especially over cities, updrafts and variable terrain could cause great uncertainty about just what dose a population might receive in any planned attack.

No leader, civil or military, considered chemical or biological use without reckoning the possible consequences of retaliation. In particular, to initiate the use of biological weapons might incite retaliation in kind, with results even more uncertain than those predicted for chemical weapons. Who knew what medieval plague might be unleashed or what new disease the enemy might have concocted?

In calculations by the major powers for both chemical and biological weapons, the responses of the irrational political actor also had to be factored. The Allies speculated whether Hitler, for example, would use chemical weapons when, in the last days of the war, his defeat was obvious and his troops were cornered. The United States brought those same speculations to estimates of what Saddam Hussein might do in the 1991 Gulf War and in the 2003 invasion. These kinds of "compound expectations," which influenced decisions about nuclear weapons, likely also figured in decisions to keep biological weapons out of theaters of war.[12]

For most of the last century, the same restraints that worked against chemical weapons use—the law and custom supported by an empowered public, technological drawbacks, widespread military disinterest, government leadership, and the reckoning of the consequences of use—have over the years reduced the risks of biological weapons, with much left to chance. As this book describes, a small number of scientists, government officials, and military officers in major powers were organizationally successful. Aided by the circumstances of war or militarized governments, they were able to begin and sustain secret biological weapons programs. Lacking political openness and review, such ventures found secure bureaucratic niches that fostered internal secrecy and avoided evaluation. In this, they were like other government programs.[13]

Scientists with visions of how the microbiology could be used for military ends were key participants in the secret offensive programs. In cases where we have records, one or two key scientists persuaded government officials to authorize the start of research and development. Then they and other scientists and engineers followed through with the crucial work in aerobiology, infectious disease research, and industrial fermentation, and even contributed to munitions design and testing.

How could biologists and physicians devote their energies to weapons patently aimed at civilians, with no other purpose than to kill life?[14] Only a few available memoirs and documents offer glimpses into the moral orders of program scientists, who, perhaps like contemporary nuclear scientists, lived within closed and unchallenged professional communities.[15] The structural similarities among programs probably gave these important participants much common experience, even if they were in enemy camps with opposing ideologies.[16]

The political context for biological weapons programs varied greatly. Democracies, dictatorships, and totalitarian governments supported offensive programs and gave their militaries latitude to expand them, even in peacetime. Always marginal, never as well funded as conventional or later nuclear weapons, these programs were extensions of existing chemical programs. With time, especially in the United States and the Soviet Union, their ambitions rivaled those of the nuclear programs in the scale of planned capacity for destruction. The Western democracies, though, were the first to renounce their offensive biological programs voluntarily. Each of them—France, the United Kingdom, and the United States—were nuclear powers that, satisfied with that one sure and certain weapon of mass destruction, abandoned the concept of future germ warfare. In contrast, the Soviet Union, having sought to maintain all strategic options, relinquished its biological weapons only as its regime was ending, as did Imperial Japan, apartheid-era South Africa, and Saddam Hussein's Iraq.

Three Historical Phases

The history of biological weapons is unwieldy in its details and riddled by missing state documentation, misinformation, and disinformation. For the purpose of narrative organization, we can say that the history of biological programs proceeded through three general phases: an offensive phase when their development and production were legitimate; a subsequent phase dominated by treaty norms prohibiting biological weapons entirely, both their programs and their use; and the current phase, still evolving, with much tension between national and international security objectives and with great issues emerging concerning public trust in government and the control of basic science for beneficent purposes.

The initial, offensive phase was characterized by programs developed in an international legal environment that allowed the development and production of biological weapons even as it condemned their use. The interpretations of reprisal, of retaliation, and of justified first use were left to the industrial states and their militaries, which moved from defensive to offensive capability (and sometimes back again) without violating any treaty. This phase began in the 1920s with the French program, the first of its kind. It focused on uniting the science of disease transmission with the new technologies of air warfare. The German occupation curtailed the French program in 1940.[17]

That same year, at war with Germany, the United Kingdom initiated its biological weapons program as a complement to its existing chemical establishment. When the United States entered the war in late 1941, it began a close cooperation with the United Kingdom and Canada that aimed at strategic capability—the power to attack behind the lines, at cities and industries—and also developed weapons for clandestine special operations. The United States brought industrial resources to the development of biological weapons when the United Kingdom had few of its own to spare.

After the war, with the Axis powers defeated, the United States, the United Kingdom, and Canada continued their preparations to wage biological warfare, while deep secrecy surrounded the Soviet Union's activities. In the Cold War era that began shortly after the war, the strategic potential of anthrax bombs was measured against atomic weapons and found promising. At this juncture, in 1947, the United States submitted its draft resolution to the United Nations defining biological weapons as weapons of mass destruction, along with nuclear weapons and lethal chemicals.

The intense secrecy that surrounded biological weapons during World War II continued during the 1950s and 1960s as the US program advanced the technologies of strategic biological weapons attacks. In this period, the United States, with help from its close allies, staged simulations of massive dispersals over American cities, over oceans, and over large regions of North America. The perceived threat was the Soviet Union and Communist China. The involvement of the United States in the Vietnam War gave more latitude to its military to develop first-use options for biological weapons than ever before.

The second phase in the history of biological weapons centered on issues of treaty compliance and noncompliance. It began in 1969, when President Richard Nixon renounced biological weapons and ended the US offensive program. This decision was followed by the Biological Weapons Convention (BWC) of 1972, which brought a new emphasis on international norms and laws. The BWC, to which 151 states are now parties, went beyond the Geneva Protocol to ban development, production, and possession. Thus:

> Each State Party to this Convention undertakes never in any circumstance to develop, produce, stockpile or otherwise acquire or retain:
>
> 1. Microbial or other biological agents, or toxins whatever their origin or method of production, of types and in quantities that have no justification for prophylactic, protective or other peaceful purposes;

2. Weapons, equipment or means of delivery designed to use such agents or toxins for hostile purposes or in armed conflict.

By this treaty, biological weapons programs were stripped of their legitimacy. Any nation pursuing an offensive program had to do so secretly or admit an overt violation of an international law and a norm accepted by most nations in the world. This total prohibition of even the development and possession of biological weapons distinguishes this second phase in the history of biological weapons from the first, in which, under the Geneva Protocol, the buildup and possession of weapons were not outlawed.

The BWC has no mandatory provision for verification (the declaration of facilities and projects, for example, and their inspection) to allow state parties to assure each other of compliance. It was written during the Cold War even before the major powers could agree on verification measures for nuclear treaties. The Soviet Union, one of the treaty's depository nations, exploited the lack of compliance provisions, creating an enormous offensive program in direct defiance of the BWC. This secret violation ultimately threatened the treaty's credibility.

After the demise of the Soviet Union in 1991, world reaction to its massive offensive program intensified international efforts to strengthen the Biological Weapons Convention. The Cold War was over; no other large industrialized power was left to defy the norm against biological weapons. The same reasoning applied to chemical weapons. In 1993, the Chemical Weapons Convention (CWC) also fortified the Geneva Protocol by banning the development, production, stockpiling, and possession of these weapons. The CWC also implemented the verified destruction of US and former Soviet chemical stockpiles. Unlike the Biological Weapons Convention, the Chemical Weapons Convention has an organizational base, located in The Hague, and three hundred professional personnel to implement compliance. At the end of the Cold War, there were hopes that the Biological Weapons Convention would develop a parallel institutional structure.

The third, defensive phase of the history of biological weapons began in the early 1990s as the Cold War ended and the United States became increasingly concerned about asymmetric warfare and bioterrorism. The emergence of a democratic government in Russia meant that the advanced industrial states had now lost all incentives, geopolitical, ideological, and economic, for total war against each other. The weapons that they had developed for strategic advantage were now proliferation liabilities in a new global context for armed conflict. According to political scientist Mary Kaldor, new wars are based on identity politics (ethnic or religious) and are waged by decentralized armies supported by globalized transfers of weapons and wealth.[18] Thus, ethnic cleansing in the 1990s was as much a phenomenon in the Balkans as in Central Africa. In these regions and elsewhere, gangster-like networks could rely on young men deracinated by shifts in global markets and flow of capital. Within this framework, Islamic fundamentalist terrorists could strike against the West, identified with individualism and other cosmopolitan values. The question then became whether states could play an effective role in preventing bioterrorism.

The extraordinary terrorist attacks of September 11, 2001, shook American complacency about domestic invulnerability to foreign aggression. The bizarre series of anonymous anthrax letters that soon followed riveted American attention on the potential danger of biological agents. The two events provoked the United States to immediate, radical response. Military action to destroy the Taliban in Afghanistan and to displace the regime of Saddam Hussein in Iraq was paralleled by the introduction of unprecedented emphasis on US civil defense. National security is about much more than the threat of biological weapons. Yet at that extraordinary time, the alleged danger of Saddam Hussein's anthrax arsenal and the potential for bioterrorists to attack American cities had a fundamental influence on making war abroad and on reorganizing the federal government for homeland security. As in 1947, germ weapons were categorized as weapons of mass destruction, although their use remained untested.

How Biological Weapons Were Invented

The offensive phase of biological warfare history was a long and, for the paths it paved, dangerous episode. The first chapter of *Biological Weapons* starts with the discoveries about disease transmission that are the foundation for biological weapons research, just as they are for microbiology and related sciences. The scientists first involved in biological weapons programs were visionaries reacting to what seemed to be dire threats to their nations or, in the case of Japan, to goals of imperial expansion. The French in the 1920s were the first innovators and so their interwar program, about which not a great deal is known, is presented here.

The early programs explored a wide variety of ways an epidemic could be deliberately inflicted: by infected insects, contaminated food or animal fodder or water, or even germ-coated feathers delivered by bombs. At the top of the list were aerosols, considered the most efficient means of infecting unprotected populations over large areas. For this, a pathogen had to be durable enough to withstand being mechanically aerosolized by a bomb blast or spray generator, either from a liquid suspension or a dry powder, without losing its virulence, that is, its capacity for infection.

Solving the problem of maintaining viable pathogens depended in part on choosing the right agent, a problem that we know American scientists investigated in detail. From the first programs until now, many of the same microbes and toxins appear on lists from the World Health Organization, the US Centers for Disease Control and Prevention, the North Atlantic Treaty Organization, and other sources. Usually around forty agents are listed. Those for anthrax, tularemia, plague, cholera, Q fever, brucellosis, and smallpox were often favored candidates for exploration in state programs. Among lethal toxins, botulinum toxin was initially the most experimented with, although it was difficult to produce in quantity and as an effective aerosol; ricin, the toxin made from castor beans, was also less efficient. The rationale for choosing disease agents also included the vulnerability of the target population. If an enemy population were likely to be protected against a disease, for example, vaccinated against smallpox, that agent might readily be abandoned. Diseases for which there were no vaccines or remedies, such as Ebola and other viral hemorrhagic fevers, were investigated as potential weapons and because they posed a threat to one's own soldiers and civilians.

The early programs were started without much proof that any pathogens could be made into effective weapons that a modern military could use with confidence. It took authoritative scientists to convince civil and military leaders that these unusual weapons could have future value. Chapter 2 describes the role of scientists in the UK biological weapons program, initiated during World War II alongside the British chemical weapons program that had been established at Porton Down in 1916. This juxtaposition allowed biologists to combine their expertise with those already familiar with aerosols, aerodynamics, bomb design, and open-air field testing. In the United Kingdom, work on the first anthrax bomb was sufficiently promising that it seemed ready for mass production.

British biological warfare scientists soon realized that they would need the resources of the United States to move from a bomb prototype to the manufacture of the half-million anthrax bombs they estimated would be necessary for initial strategic attacks on Germany. Chapter 3 reviews how the wartime United States took on the challenge of biological weapons research and production, enveloping it in great secrecy. In the three years in which it operated, it gave its leaders the resources to explore both defensive and offensive options. After the war, its leaders could rightly claim, although not in public, they had brought biological weapons to an industrial level of production.

Chapter 4 turns to the biological warfare program of the Japanese Imperial Army. In 1947, US biological weapons experts became involved with US intelligence in arranging an immunity bargain with Japanese scientists in exchange for information about their experiments with and use of biological agents. Those experiments included vivisection and other grotesque torture of Chinese captives. This secret bargain freed the Japanese scientists involved from war crimes prosecution and helped set the stage for the next two decades of the secret American program and its strategic mission.

Chapter 5 traces the expansion of the US program following the Japanese bargain, when American scientists struggled with persuading the armed services that biological weapons could be predictable, potent strategic weapons. In 1951, North Korea accused the United States of using Japanese-style biological weapons, mainly insects, against its civilians. The context was the Korean War, and this allegation became the first Cold War example of the difficulties of sorting evidence from ideology and establishing the facts behind accusations. The West defended the United States, while North Korea, China, and the Soviet Union held with their belief that Americans had, like the Japanese, crossed the line from production to use.

Assimilation by the military depended on field demonstrations of entire weapons systems. Within a decade, the US program's field trials became extensive, ranging over all conceivable terrains, from the Arctic to tropical jungles, and they were backed up by human-subjects research that aided in understanding dose response and in standardizing tularemia weapons. With the latitude afforded the military during the Vietnam War, biological weapons verged on the military assimilation their advocates sought. Few people in government or in the public were aware that these new strategic weapons were being developed. Instead, popular opinion was against chemical weapons, especially tear gas and herbicide use in Vietnam, as illegal threats to human life and the environment. The enormity of the biological program remained in the shadows.

The Second Phase: Normative Compliance

Following a comprehensive review of the US biological program in 1969, President Richard Nixon decided to renounce biological warfare. This decision introduced a twenty-year period that emphasized legal norms as a means of biological arms control. Nixon fostered the long-delayed US ratification of the 1925 Geneva Protocol and promoted the 1972 Biological Weapons Convention. Chapter 6 reviews Nixon's role and that of Henry Kissinger, his national security advisor, as well as the roles of other crucial actors—scientists, politicians, and journalists—who contributed to the momentum to stop proliferation. During this long period, the focus on biological weapons as "weapons of mass destruction" virtually disappeared.

With no effective process to promote transparency, the Cold War era was characterized by charges levied by one nation against another. Ideological divides also allowed a blind eye to be turned on the suspect activities of allies and political pawns. Chapter 7 presents an overview of the Soviet program, still poorly documented, from its early inception to the 1992 declaration by Russian President Boris Yeltsin that violations of the BWC had occurred.

Chapter 7 also covers the two major accusations about the use of biological weapons made by the United States against the USSR. In both, independent scientists stepped into the breach to compensate for the lack of BWC verification procedures. The first case was the 1981 US allegation that the Soviets had conspired to use mycotoxin agents (called "yellow rain") in Southeast Asia. The second controversy concerned the 1979 anthrax epidemic in the closed Soviet industrial city of Sverdlovsk. Both allegations, as with those made against the United States during the Korean War, pitted ideology against evidence and state secrecy against transparency. These same dilemmas about evidence gathering would surface again, with special regard to Iraq and its program but also, in theory, to any nation rejecting openness and compliance with the Biological Weapons Convention.

The Third Phase: Defense

The third and current phase in the history of biological weapons is one of defensive response, with the United States as the world's lone superpower taking the lead in defining threats and how to contend with them. Chapter 8 considers bioterrorism, the new vision of the threat of biological weapons that emerged during the 1990s and especially in President Bill Clinton's second term.

During that time, international negotiations continued for a BWC protocol that would strengthen compliance measures. Simultaneously, federal officials made efforts to heighten public consciousness of bioterrorism and the need for government response, apart from intelligence gathering and other means for combating terrorism in general. The basis was established for national "domestic preparedness" programs to protect large cities. The foreboding was not only that Americans would be targeted by terrorism as indiscriminate in its purposes as total war, but also that terror would take the form of disease outbreaks.[19]

Meanwhile, along with revelations about the Soviet program, the world learned about the offensive biological program in Iraq and about another program, in apartheid South Africa. This chapter considers both these programs as examples of weapons proliferation to lesser states. The way in which evidence about each program was obtained, one through United Nations inspections and the other through the courts and public hearings, offer insights on two unusual processes that increased civilian safety. In Iraq, enforced transparency spurred the elimination of Saddam Hussein's nuclear, chemical, and biological weapons. In this period, especially concerning Iraq, one hears biological weapons again identified as weapons of mass destruction, along with chemical weapons and, of course, nuclear arms.

In chapter 9 we review the effects on policy of the Al Qaeda attacks of September 11, 2001, and the anonymous anthrax letters that had their greatest impact in October and November of that year. For a time, it seemed anthrax spores were everywhere, disrupting the mail and causing general anxiety. The anthrax letters infected twenty-two people in Florida, New York, New Jersey, Connecticut, and the District of Columbia, and killed five of them, none of them the targets of the letters.

Anthrax was also an issue in how the United States and the United Kingdom construed the threat of Iraq. When Secretary of State Colin Powell argued the case for the invasion of Iraq to the UN Security Council, he cited Saddam's anthrax weapons in his list of justifications. In only two years, as chapter 9 reviews, President George W. Bush used a combination of military force, federal reorganization for civil defense, and technology-driven policies to combat the perceived threat of bioterrorism. The problem has become whether a national security policy alone, with its increased secrecy and centralized authority, best protects citizens against the risks of biological attack. How can the public trust government interpretations of risk when they seem politically constructed from a haze of uncertain intelligence?

The world is much different now from what it was for most of the twentieth century, when it was easier for Americans to retreat from global issues. The question most relevant to our present subject is whether the world is better protected now against biological weapons than it was then. On the positive side, the major powers, and indeed almost all nations, are genuinely committed to the treaties outlawing these weapons, which was not true until the Cold War ended. Hiding a biological program is probably more difficult now, with improved surveillance technology and more open travel and communication. And few governments are left whose populations want the ostracism and risks associated with biological weapons.

Yet the problem of biological weapons proliferation may be greater now than ever. Advances in biotechnology could be used to destroy life as well as to modify basic processes such as cognition, development, reproduction, and genetic inheritance. These innovations, which are fast accelerating to further medical science, also offer "unprecedented opportunities for violence, coercion, repression, or subjugation."[20] As in the past, the most dangerous factor is government secrecy that hides from the nation and the world covert attempts to develop and use biological weapons. Can science be controlled? Can there be international accord sufficient to avoid a catastrophic new generation of biological weapons programs?

The final chapter of this book takes the approach that no single restraint can successfully keep biological weapons from proliferation or use. A range of proposed solutions are now available, from legal alternatives, trade controls, and federal and local homeland security initiatives to the recruitment of microbiologists for biodefense. This new area of policy is lively and still taking shape. Many times during the twentieth century, one could point to thousands of people in one or another country busily engaged in an offensive biological weapons program. Now tens of thousands of people in the United States are committed to protecting their society against such weapons and against bioterrorism, and other governments are similarly alerted. These efforts need to be balanced with stronger international networks that secure the legal means for stopping proliferation before it starts, with positive and negative sanctions. Overall, the change in norms from the last century to this one should make a difference, not simply to promote exercises in national defense, but also to provide enduring protection for all civilians against biological weapons.

CHAPTER 1

Biological Agents and Disease Transmission

For most of history, people have believed that the transmission of disease is a mysterious phenomenon controlled not by humans but by the gods, witchcraft, or fate. Collective disease, including the sudden epidemics that decimated cities and armies, was a frequent but misunderstood occurrence. Often relying on the ancient Greeks, Western scholars and physicians from the Middle Ages well into the nineteenth century held firm to the notion that "miasma," the stench of putrefaction, was the source of many epidemics and that changes in the weather or the planets increased the chances of outbreaks.[1] One of the positive results of the belief in miasma was the instigation of campaigns to clear cities of garbage, open sewage, standing water, slums, and unhygienic slaughterhouses, measures that significantly reduced the risks of infectious diseases. In both Europe and Asia, public health measures allowed urban economic growth. But whole societies remained vulnerable to devastating epidemics that were explained either fatalistically or by a science still struggling for proof.

Toward the end of the nineteenth century, before any state biological weapons program, medical scientists discovered microorganisms and made great strides toward understanding that a specific germ can cause a specific disease; that food, water, and personal contact communicate illness; that a pathogen can cycle through different species; and that insects and protozoans play a role in creating epidemics. Once these causal links were understood, humans could methodically control disease outbreaks. It became more possible to protect populations from the great assaults of plague, cholera, diphtheria, smallpox, influenza, and malaria that had swept across nations in previous centuries, hitting hardest in crowded urban centers and among the poor. Scientific knowledge by itself was no magic wand. Wars, forced migrations, famines, malnourishment, preexisting illness, and great poverty, especially in the colonial empires, remained the political preconditions of epidemics that science alone could not address then any more than it can now. Together, stability of life, public health campaigns, and scientific knowledge about disease transmission aided human survival.

By the 1920s, Western societies were rarely susceptible to sudden deadly outbreaks of the sort that threatened social order. Public health in cities had improved, water and food sources were monitored by the state, and vaccines and drug therapies were being invented as further protections. With most childhood diseases conquered, more people were living longer, a trend that has continued, and they were dying of the diseases of industrial society that afflict older people, such as cancer, heart disease, and stroke. In the rest of the world, scientific knowledge alone without public health and freedom from war and poverty failed to prevent large epidemics. The enduring dichotomy between developed versus developing nations, or North versus South, remains marked by generally good health versus widespread, preventable epidemics.

As Western nations were distancing themselves from the collective catastrophes of epidemics, some of their governments invented biological weapons as a means of achieving advantage in warfare. The German military during World War I made the first foray—against animals, not people. It mounted an international sabotage campaign to kill packhorses and mules that were being shipped to the British and French from neutral nations, including the United States, Norway, Spain, and Romania, and from South American ports.[2] Stevedores were paid to infect individual animals with anthrax or glanders, and entire shiploads of animals were sickened and killed. From the German perspective, these attacks violated no international norm.[3] For years, though, this innovation cast suspicion on what further covert use of biological agents Germany might devise.

This chapter begins with the establishment of germ theory and then proceeds to an account of the early days of biological weapons, when the French sought to best the Germans by integrating aerosols from bombs with the new potential of air warfare. The tension between preparation for reprisal against an enemy nation and prohibitions against biological weapons characterize this early episode in biological weapons history. At the same time as the French were signing the 1925 Geneva Protocol, they were developing a biological warfare program to complement the one they had established for chemical weapons during World War I.

Little is known about how scientists first approached germ weapons. Included in this chapter, therefore, is a review of a remarkable early document, "Bacterial Warfare." It was written in 1942 by two American biologists, Theodor Rosebury and Elvin A. Kabat, soon after the Japanese attack on Pearl Harbor, when scientists were joining up for the war effort. Classified at the time, the report was published openly in 1947.[4] Rosebury and Kabat and others like them stood on the brink of an entirely new endeavor, on which they were willing to embark as patriots. The report's summary of the potentials of biological weapons and how to defend against them still resonates today, more than sixty years after it was written. The statement also expresses appropriate moral hesitation about the use of biological weapons against civilians, balanced against the deterrent effect such weapons might have on the enemy.

Epidemics

Throughout the recorded history of epidemics, ignorance of disease transmission has increased the risks of death and illness. Without knowing the source of an outbreak or the cause of its spread, large populations have been left defenseless.[5] The great cholera pandemic that devastated Europe in the 1830s offers a catastrophic scenario that microbiology could later help prevent, if governments acted openly to educate the public and maintained basic public health services. The outbreak began in India in 1817, and, in a second wave in 1826, it spread to Moscow and from there into Western Europe. No one then understood that a bacterium, the cholera *vibrio,* causes the disease or that it is communicated largely through feces-contaminated drinking water. The disease reportedly killed 40 to 70 percent of those infected in a few days or even a few hours. Militarily enforced quarantines of infected communities caused riots and violence.[6] Throughout the cholera epidemic, European doctors explained that it was caused by "atmospheric conditions" and miasma or noxious stench.[7]

In the 1800s, physicians had difficulty distinguishing one epidemic disease from another. Fevers, rashes, loss of appetite, and aching joints were symptoms common to half a dozen likely "plagues."[8] One exception to this diagnostic confusion was smallpox, with its distinctive rash, for which a preventive inoculation was possible, although not everywhere accepted. For centuries itinerant folk doctors in China and India used the scabs from infected victims to inoculate others against the disease. This practice was communicated via the trade route to Constantinople, where, in 1720, Lady Mary

Wortley Montagu, wife of the British ambassador, discovered variolation (as it is now called, from the variola virus) and brought it back to the British court. This innovation caused a theological and medical controversy about deliberate poisoning of the body.[9] In 1796, investigating why farming people were often unscathed by smallpox, the English physician Edward Jenner conducted experiments on inoculation and then developed a serum from cows that could be injected to prevent smallpox infection. The medical world soon became divided between those who accepted Jenner's method and those who doubted or feared it. Another hundred years would pass before immunization and vaccination were understood. Vaccinations figure strongly in biological weapons history, as they do in military and public health history, for the protective benefits they confer and for the anxieties they continued to provoke about risky side effects and bodily pollution.

In the last half of the nineteenth century, European scientists approached the study of disease as a quest for the secret laws of nature, which persistence, inspiration, a good microscope, and a rudimentary laboratory for experiments might reveal. In 1858, French physician Louis Pasteur published his argument that specific germs cause disease, which he based on fermentation experiments. Then, setting out to disprove the theory that germs generate spontaneously in organic matter, Pasteur showed that germs are actually invisibly airborne. In 1876, Pasteur's illustrious rival, the German physician Robert Koch, gave rigorous proof to the germ theory by his experiments with *Bacillus anthracis,* the same anthrax-causing bacteria that became a preferred biological weapons agent in the next century. Koch used his pure culture technique to track the life cycle of the anthrax bacterium, from its dormant spore form through its germination to a rod-shaped growing form and back again to the spore. Koch went on to discover the tubercle bacillus that causes tuberculosis and the cholera *vibrio,* and he set the laboratory standards for isolating and growing bacteria. Meanwhile, Pasteur invented vaccines, which he named from the Latin word for cow, *vacca,* in deference to Jenner's work. His anthrax vaccine, announced a success in 1881, helped end livestock outbreaks (epizootics) throughout the world.

Germ theory, like Lady Montagu's smallpox inoculation and Jenner's serum, was resisted at first by the medical establishment. Those experimental scientists who understood its importance began using new basic laboratory techniques to culture disease agents and to investigate vectorborne diseases. Western microbiology, building on the remarkable French and German advances, became the foundation for all biological warfare programs in the next century. It was by no means obvious that microbes could be used as agents for weapons. As far as was known, they were not stable outside a host. Unlike chemical compounds, they seemed too fragile to use as bomb fill, which would subject them to the mechanical shock of discharge and high temperature. Both of these qualities, stability and hardiness, had to be scrutinized before biological weapons could be thought of as feasible.

The French Biological Weapons Program

In 1919, at the request of the French government, Auguste Trillat, director of the Naval Chemical Research Laboratory, conducted an inspection of a German pharmaceutical plant, as part of the oversight instituted by the terms of the Treaty of Versailles and the Inter-Allied Control Commission. At the plant, the director of the bacteriological laboratories apparently confided that German research on biological weapons was continuing.[10] The French were well aware of Germany's wartime campaign against pack animals, and France itself may have engaged in similar sabotage, with undercover agents and prisoners of war using glanders or anthrax against German livestock.[11] Trillat's report back to the French government raised alarm. The Germans had their fears as well, stimulated by rumors about Soviet bacteriological bombs. But their military experts were skeptical about biological warfare, advising that it might backfire on the attacker, and they focused instead on chemical weapons.[12] In 1921 the French War Ministry concluded that France should begin a biological weapons program. To lead their program, the military turned to Trillat, who had already conducted experiments on the airborne transmission of bacteria.

Trillat was probably the first government scientist to calculate the potential military value of biological weapons and to play a role in implementing them. He believed that liquid cultures loaded into shells and bombs could be detonated to form "microbial clouds" with great infective power.[13] Furthermore, he convinced French authorities that the most efficient means of attack would be biological bombs dropped from aircraft.[14] The future use of airplanes for chemical bombing was simultaneously envisioned by others, among them the French military leader Marshal Foch, who wrote in 1921, "The carrying power of the aeroplane is increasing. Improvements are made almost daily enabling greater and greater weights to be carried. These developments introduce an entirely new method for the large-scale use of poison gas. By the use of bombs, which are becoming increasingly efficient and of greater capacity, not only have armies become more vulnerable, but the centres of population seated in the rear, and whole regions inhabited by civilians will be threatened."[15]

Trillat had limited confidence in technology for protecting soldiers against germ weapons, for example, via gas masks; nor did he believe that vaccines would offer full protection. He saw biological weapons doing the most harm behind enemy lines—against reserve troops or against civilians in industries and cities, and against livestock, crops, and water supplies. To build retaliatory power as a defense, he advocated research on anthrax, brucellosis, cholera, dysentery, plague, and typhoid as potential weapons.

Making a pathogenic aerosol was a challenge. Like many biologists at the time, Trillat believed that microbes were fragile but could be sustained outside a host with moisture, the way droplets from coughing or sneezing spread germs. His experiments confirmed his theory that slurries of agents best kept their virulence; he therefore rejected the idea that pathogens might be dried and efficiently disseminated, which scientists later proved possible.[16]

In discussions then about the drawbacks of germ weapons, scientists doubted that microbes could withstand explosive impact.[17] Trillat, however, believed that the shock of an instantaneous explosion would disperse microbes without destroying them, so that an infective biological aerosol from bombs was feasible.[18] Trillat also saw the air as a "vast field of culture" that, provided it was humid, would encourage the growth of microbes and allow the effective dispersal of germs over a city. His approach was more than poetic, for he had studied variable atmospheric conditions and their impact on the virulence of microbial aerosols.

How far the French went with their biological warfare plans remains unknown, owing to the disruptions when Germany occupied France in 1940 and to French secrecy both before and after the war. Pasteur's institutional legacy gave France a strong system of national and international biological research institutes, which, along with scientific training in its universities and medical schools, provided a decided advantage in exploring biological weapons. In addition to research at the Central Naval Artillery Laboratory, aerosol and bomb trials were conducted at Le Bouchet National Explosives Plant, and animal experiments were conducted at the army's Veterinary Research Laboratory in Paris. The substance of this research is difficult to evaluate, except through a few publications by Trillat and his colleagues. It seems that little was known even to France's allies, although the British and French cooperated before the war on testing chemical weapons in Algeria.[19]

For a period between 1927 and 1934, the French biological warfare effort appeared to diminish, most likely owing to France's commitment to the Geneva Protocol. Then the French perception of a German threat intensified after Hitler came to power, with particular fears about German air attacks.[20]

French anxieties were spurred in part by allegations in the press that German spies had been using the Paris Metro to test the airborne dispersion of biological agents. French scientists, in reviewing German literature, also found an unsettling optimism about the feasibility of biological bombs. One German source was quoted as saying "Victorious will be the nation that knows how to find the most virulent bacillus and the most effective vaccine for defending itself against it."[21] At a meeting of French military personnel in 1937, the argument was made that the Germans in war would be interested in a general conflagration and therefore would be ingenious in mounting a full-scale war "to assure the destruction of soldiers and of civilians behind front lines, no matter what their age or sex."[22]

Between 1935 and 1940, the French military was again actively researching biological weapons, which was not the case in the United Kingdom, the United States, Germany, or any other Western power. By this time, logistical problems were coming into sharper focus. The rationale for choosing among potential pathogens still had to be decided. What was the target—humans, animals, crops, or water? If human, the condition of the host population mattered; this meant whether its immunity was already weakened by the conditions of war, bad weather (especially the cold), fatigue, or demoralization, and whether age, sickness, or race might affect vulnerability. What was the preferred route of infection? Aerosols were considered most efficient; bacteria might also be introduced by a cut or insect bite or by food or water contamination. The pathogen had to be virulent, that is, capable of rapidly causing serious disease. Could it sustain its virulence over time, through the process of fermentation and after being loaded into munitions and stored? Would the munitions efficiently generate aerosol without inactivating the pathogen? In addressing these and other questions, the French prefigured the approaches of scientists in the later state programs. So, too, did their first choices of possible agents, for the diseases tularemia, brucellosis, meloidosis, plague, and anthrax. They also considered toxins produced by bacteria, especially botulinum toxin (produced by the bacillus *Clostridium botulinum*), a favorite in the later state programs.

In the open French literature, arguments also appeared for civil defense against a possible biological warfare attack. One, from a military physician, advocated a new organization (DCM, Defense Contre Microbes) and proposed three tactics: (1) heightened detection of pathogens by testing air, dust, water, and food and by capturing and testing rats in cities and ports; (2) individual protection of citizens with face masks and vaccines, and (3) the spraying of prophylactic "clouds" of antiseptics in buildings and other inhabited spaces, an idea earlier proposed by Trillat.[23]

In the spring of 1940, as German troops advanced, nearly all French documents at the various research stations were reportedly destroyed or hidden and the program evaporated. Despite British fears that French biologists at one of the Pasteur Institutes might cooperate with the Germans on biological warfare experiments, no evidence for collaboration surfaced.

While the French took seriously the perceived German biological warfare threats, American military experts reacted with skepticism. The defenses available to counteract germs seemed greater than any danger of biological attack. In 1933, for example, Maj. Leon Fox wrote an article in *The Military Surgeon* in which he reviewed the possibilities of biological warfare, proceeding from intestinal disease to respiratory illness to insect-transmitted outbreaks. In the first category, he trusted modern sanitary measures that reduced the chances of contaminated water, milk, and food. As for respiratory diseases, he put his faith in both natural immunity and the fact that the worst of these (to his mind, influenza, pneumonia, and epidemic meningitis) were contagious and would rebound as badly on the side using them as on an adversary. Fox made an exception for anthrax (which is not contagious from person to person), but he did not consider it a serious threat. As for plague and typhus, the insectborne diseases he considered most important, Fox pointed to the relative ease with which fleas, lice, and rodent reservoirs can be destroyed to contain any outbreaks. Furthermore, he believed that pragmatic consideration alone would determine whether biological or, for that matter, chemical weapons would be used in future wars. He was pessimistic that biological weapons would make the grade, noting that "at the present time practically insurmountable technical difficulties prevent the use of biologic agents as effective weapons of warfare."[24]

Rosebury and Kabat on Biological Weapons

Within a decade, as war broke out, biologists in the United Kingdom and the United States took on the "insurmountable technical difficulties" as challenges. In 1942, Theodor Rosebury and Elvin Kabat, from Columbia University's College of Physicians and Surgeons, mapped the components of a comprehensive biological weapons program—from choosing biological weapon agents to framing offensive and defensive objectives.[25] In a year's time, with assistance from the British, the United States started its biological warfare program. Rosebury joined it as a division chief and served to the end of the war; his junior colleague Kabat also contributed research to the program.

Some fifty pages long, the Rosebury-Kabat report begins with instructive comparisons between chemical and biological weapons. Like Trillat, whose work they knew, Rosebury and Kabat observed that both chemical and biological weapons agents, invisible, intangible, and diffused over large areas, might have psychologically frightening (what they termed "insidious") effects on civilians. In addition, chemicals and pathogens can equally poison the body and cause sickness and death. What, though, were the distinctive characteristics of biological agents? The authors noted six features:

1. Infective biological agents have *incubation periods.* That is, their effects are delayed and can take days to appear.
2. Disease agents can also be *contagious.* That is, they can spread from one human to another, and they can be spread by animals and by contaminated food and water.
3. Disease agents are characterized by *infectivity,* the frequency with which they can induce symptoms among individuals in a group.
4. Infectivity, though, can vary according to individual *immunity.* Individuals differ in their susceptibility to contracting a given disease.
5. Infective agents, unlike chemical agents, are not manufactured. They are life forms that *proliferate.*
6. Without their characteristic *mammal* host, disease agents tend to be *unstable* and lose their virulence, although virulence can be technically enhanced.

Each of these features spoke to technical problems that Trillat had confronted and with which the British and Americans would also contend. Much about the disease agents selected for study was unknown. The actual impact of any intentional epidemic was theoretical and would remain so.

Rosebury and Kabat found Trillat's preoccupation with slurries and humid atmospheric conditions less interesting than the transmission of germs through the air. They turned to research indicating that bacteria sprayed into the air can be recovered from surfaces hours afterward with no loss of virulence; the agents of pneumonia, cowpox, influenza, polio, and tuberculosis had more stability than once believed. Along with reported experiments on stability, Rosebury and Kabat used the literature on laboratory accidents to inform their discussion. For instance, a mysterious outbreak of brucellosis at Michigan State University was traced to aerosols; dozens of laboratory mishaps involving tularemia showed the sturdiness of microbes dispersed over distance and left undisturbed over time. These two scientists knew more about drying techniques, such as freeze-drying (lyophilization), than Trillat had in earlier days. All this knowledge pointed to a potentially successful integration of pathogens into modern warfare.

What would be the best delivery system for a biological agent? "The airplane," Rosebury and Kabat concluded, "is clearly the most useful means for the dissemination of infective agents."[26] But the authors refrained from endorsing bombs or shells. Despite Trillat's confidence, much was still unknown about the dispersal and survival of bacteria in explosions. A low-flying airplane might, they thought, directly assault ground troops or civilians with biological weapons using vectors such as infected fleas to spread plague, as the Japanese were accused of using in Manchuria.[27] Airplanes might potentially pollute American water reservoirs. Furthermore, the dissemination of airborne agents needed experiments, for example, about which freeze-dried agents might best lend themselves to infective clouds.

From the outset, the authors were against any first use of bacteriological weapons. They believed instead in the investigation of biological weapons for defensive purposes and in their development for deterrence. In their report they cited the 1942 radio speech by British Prime Minister Winston Churchill declaring that, should the Germans resort to poison gas, the British would retaliate "on the largest possible scale far and wide against military objectives in Germany."[28] Rosebury and Kabat responded, "The likelihood that bacterial warfare will be used against us will surely be increased if an enemy suspects that we are unprepared to meet it and to return blow for blow."[29]

Starting anew, Rosebury and Kabat imagined the requirements of the entire biological weapons production scheme. If the military intended industrial-level production, they pointed out, it would have to build or convert plants comparable to or larger than commercial vaccine factories. Highly infective agents such as those for plague or tularemia would need an entire isolated building. If the military developed highly virulent or drug-resistant variants, it should be on the alert for the problems of disease transmission that accidents with unusual strains could cause.

The Candidate Pathogens

Most of the Rosebury-Kabat report is devoted to studious review of seventy likely agents. The authors rejected thirty-seven and proposed thirty-three for detailed consideration. By comparing their list to contemporary biological warfare agent lists (table 1.1), we can see that most of the agents remain. Bacteria like anthrax and tularemia, and a few viruses, notably smallpox and yellow fever, are relatively constant candidates, although it should be noted that anthrax at this time in the US was still considered more as the cause of epizootics (animal outbreaks) than as a hazard to humans, among whom cases were rare. A range of hemorrhagic fever viruses represent additions from after World War II, for example, Lassa fever, Marburg fever, and Ebola.

The two scientists scrupulously evaluated the chosen microbial candidates, using ten criteria that ranged from the agent's availability (whether it could be easily cultivated) to the possibilities of specific immunization and effective therapy—defenses an enemy might easily have or acquire. Other important criteria concerned what percent of an exposed population might likely fall ill or die and how quickly, if the disease were contagious and about all its possible modes of transmission, and if a dispersed agent might contaminate the environment or pose a retroactive danger to friendly troops as they attempted to occupy an area.

The goals of total war determined the selection of potential agents, with no necessary distinction between causing intentional epidemics among civilians and destroying their food sources. Thus, the military purposes of biological weapons were envisioned primarily "for the disorganization of industrial areas behind the lines or of army centers and camps; for use as a part of a 'scorched earth' policy, and against valuable animals, food plants and industrial crops."[30] The siege of cities using biological agents and how cities might break such a siege were also briefly considered.

Some well-known disease threats failed to qualify. Leprosy, for example, had too long an incubation period. Diphtheria, understood as a children's disease, had too low an "attack rate" in adults, that is, too few victims would be badly infected. Smallpox and tetanus presented the problem of an enemy's being vaccinated or immune from previous exposure. The virus for poliomyelitis cultivated poorly and had low infectivity rates. Tuberculosis was a chronic disease with "low casualty effectiveness."

Botulinum toxin was first on the Rosebury-Kabat list. It was described as the most potent of all gastrointestinal poisons, by a wide margin, deadly in small quantities and with a short incubation period. Smallpox failed to make the candidate list; vaccination would make it ineffective. The list also excluded cholera and typhus, which later interested US and other state programs. On the list were influenza and Weil's disease (leptospirosis), a usually nonfatal waterborne bacterial infection, neither of which generated much subsequent military interest. Dysentery (shigellosis), a scourge in previous wars, ranked low on this list (and others) because normal sanitary precautions could defeat it. Rosebury and Kabat's concern that malaria might be introduced by the purposeful release of mosquitoes, an idea floated by the

TABLE 1.1 Biological Agents Cited as Possible Weapons for Use Against Humans

Biological agent and WHO numeric code for the disease[a] it can cause	United Nations[b] (1969)	WHO[c] (1970)	BWC[d] CBM-F (1992)	Australia Group[e] (1992)	NATO[f] (1996)	CDC[g] category A (2000)	BWC[h] draft Protocol (2001)
BACTERIA (INCLUDING RICKETTSIA AND CHLAMYDIA)							
Bacillus anthracis, A22 (anthrax)	X	X	X	X	X	X	X
Bartonella quintana, A79.0 (trench fever)				X			
Brucella species, A23 (brucellosis)	X	X	X	X	X		X
Burkholderia mallei, A24.0 (glanders)	X	X	X	X			X
Burkholderia pseudomallei, A24 (melioidosis)	X	X	X	X	X		X
Franciscella tularensis, A21 (tularaemia)	X	X	X	X	X	X	X
Salmonella typhi, A01.0 (typhoid fever)	X	X		X	X		
Shigella species, A03 (shigellosis)	X				X		
Vibrio cholerae, A00 (cholera)	X	X		X	X		
Yersinia pestis, A20 (plague)	X	X	X	X	X	X	X
Coxiella burnetii, A78 (Q fever)	X	X	X	X	X		X
Orientia tsutsugamushi, A75.3 (scrub typhus)					X		
Rickettsia prowazekii, A75 (typhus fever)	X	X	X	X	X		X
Rickettsia rickettsii, A77.0 (Rocky Mountain spotted fever)	X	X		X	X		X
Chlamydia psittaci, A70 (psittacosis)	X				X		
FUNGI							
Coccidioides immitis, B38 (coccidioidomycosis)	X	X			X		

TABLE 1.1 Biological Agents Cited as Possible Weapons for Use Against Humans

Biological agent and WHO numeric code for the disease[a] it can cause	United Nations[b] (1969)	WHO[c] (1970)	BWC[d] CBM-F (1992)	Australia Group[e] (1992)	NATO[f] (1996)	CDC[g] category A (2000)	BWC[h] draft Protocol (2001)
VIRUSES							
Hantaan/Korean haemorrhagic fever, etc, A98.5		X		X	X		
Sin nombre, J12.8							X
Crimean-Congo haemorrhagic fever, A98.0		X		X	X		X
Rift Valley fever, A92.4		X		X	X		X
Ebola virus disease, A98.3				X	X	X	X
Marburg virus disease, A98.4		X		X		X	X
Lymphocytic choriomeningitis, A87.2				X			
Junin, A96.0 (Argentine hemorrhagic fever)				X	X	X	X
Machupo, A96.1 (Bolivian hemorrhagic fever)				X	X		X
Lassa fever, A96.2				X	X	X	X
Tick-borne encephalitis/Russian spring-summer encephalitis, A84.0/ A84	X	X		X	X		X
Dengue, A90/91	X	X		X	X		
Yellow fever, A95	X	X		X	X		X
Omsk hemorrhagic fever, A98.1					X		
Japanese encephalitis, A83.0		X		X			
Western equine encephalomyelitis, A83.1		X		X			X
Eastern equine encephalomyelitis, A83.2	X	X		X	X		X
Chikungunya, A92.0	X	X		X	X		
O'nyong-nyong, A92.1		X					
Venezuelan equine encephalomyelitis, A92.2	X	X	X	X	X		X
Variola major, B03 (smallpox)	X	X		X	X	X	X
Monkey pox, B04				X			X

TABLE 1.1 Biological Agents Cited as Possible Weapons for Use Against Humans

Biological agent and WHO numeric code for the disease[a] it can cause	United Nations[b] (1969)	WHO[c] (1970)	BWC[d] CBM-F (1992)	Australia Group[e] (1992)	NATO[f] (1996)	CDC[g] category A (2000)	BWC[h] draft Protocol (2001)
White pox (a variant of variola virus)				X			
Influenza, J10,11	X	X			X		
PROTOZOA							
Naeglaeria fowleri, B60.2 (naegleriasis)							X
Toxoplasma gondii, B58 (toxoplasmosis)		X					
Schistosoma species, B65 (schistosomiasis)		X					

Source: *Public Health Response to Biological and Chemical Weapons: WHO Guidance, 2004*

Notes

[a] Diseases are identified by the numeric code assigned by the WHO *International Classification of Diseases*, 10th ed.

[b] United Nations, *Chemical and Bacteriological (Biological) Weapons and the Effects of Their Possible Use: Report of the Secretary-General* (New York, 1969).

[c] World Health Organization, *Health Aspects of Chemical and Biological Weapons: Report of a WHO Group of Consultants* (Geneva, 1970).

[d] UN Office of Disarmament Affairs, compilation of declarations of information by BWC states parties in accordance with the extended confidence-building measures agreed at the Third Review Conference, DDA/4–92/BW3 plus Add. 1, Add. 2, and Add. 3, data from Section 2, *Past Offensive Biological R&D Programmes*, of Form F as filed by Canada, France, Russia, the United Kingdom, and the United States in 1992.

[e] Australia Group document AG/Dec92/BW/Chair/30 dated June 1992.

[f] *NATO Handbook on the Medical Aspects of NBC Defensive Operations*, AmedP-6(B), Part II—Biological, 1996.

[g] Centers for Disease Control and Prevention: Biological and Chemical Terrorism: Strategic Plan for Preparedness and Response. Recommendations of the CDC Strategic Planning Workgroup, *Morbidity and Mortality Weekly Report* 49, No. RR-4 (2000): 1–14.

[h] Ad Hoc Group of the States Parties to the Convention on the Prohibition, Development, Production and Stockpiling of Bacteriological (Biological) and Toxin Weapons and on their Destruction, document BWC/AD HOC GROUP/56–2, at 465–466, which is in Annex A of the Chairman's Composite Text for the BWC Protocol.

French, was not generally shared, although the United States did extensive defensive and offensive research on mosquito-borne diseases in general during the war and later.

The Select Agents

The consensus about potential biological agents has remained high for decades.[31] Rosebury and Kabat varied only a little in their judgments from later assessments. They ranked *Bacillus anthracis* in its dormant spore form as overall the most important agent, noting that it "surpassed by few microorganisms in infectivity for animals, and by none in host range."[32]

Most of the information about anthrax infection among humans came from industry. A worker in a woolen mill, for example, might contract either inhalational anthrax or cutaneous anthrax, through a skin abrasion or cut. In areas of the world without reliable animal vaccination and veterinary inspection, infected meat could cause intestinal anthrax, also highly lethal. There was a theory, dating back to Pasteur, that abrasions in the respiratory tract aided anthrax infection and that inhalation anthrax itself, also called "woolsorter's disease," was associated with large amounts of infected dust, a "mechanical irritation factor" that increased the chances of infection. Because there was no fully reliable therapy, anthrax was and remained too dangerous for human experiments to test such theories. Rosebury and Kabat understood that untreated inhalational anthrax was associated with high fatality rates. They believed its incubation period to be short, from a few hours to several days; although this was true for most victims, longer incubation periods for some people were later suggested by data from animal experiments and by the 1979 Sverdlovsk outbreak.[33]

From a weapons perspective, the hardiness of anthrax spores was their outstanding characteristic. They lasted in the environment for years in cold and heat without losing virulence; in the laboratory, they could tolerate up to 100 degrees centigrade when dry. As for protection, the authors noted a problem: Allied troops would have to be vaccinated but no vaccination was prepared, except for animals. Even for animals, vaccinations had negative effects, with as much as one percent fatalities in herds. Well ahead of their times, Rosebury and Kabat called for the development of "a completely innocuous but effective [anthrax] vaccine." Decontamination of anthrax spores after an attack, they predicted, would present great difficulties.

Plague was ranked high as a disease candidate. In human history, bubonic plague (referring to the buboes or swellings in the armpits and groin) has caused devastating epidemics spread by infected fleas hosted by rats. If the infection in humans spread to the lungs, pneumonic plague, with higher fatality rates, sometimes followed from bubonic plague and spread through close person-to-person contact. The plague bacillus proved so stable in laboratory experiments that freeze-drying and dispersing it as an aerosol—to create pneumonic plague—seemed possible. "There is no reason to doubt that virulent plague bacilli could be disseminated by the airborne route, and that under conditions which can be rather clearly defined, a devastating epidemic could result."[34]

High on the Rosebury-Kabat list, agents for tularemia and brucellosis later emerged as important in biological weapons programs. After the war, years of US human subjects research on the dose response for tularemia (also called rabbit fever) allowed its agents to become standardized for munitions fill, along with anthrax and yellow fever.

Brucellosis has several recognized varieties that can infect animals (especially livestock) and humans. For humans, *Brucella melitensis* causes the most severe symptoms. The disease, with an incubation period anywhere between four days and four months, can last as long as three months with possible later recurrences. With a low death rate, estimated as less than 2 percent, brucellosis would figure in later programs as a "humane" alternative to more deadly diseases. In addition, the virus for psittacosis (parrot fever) was judged potentially "one of the most useful agents of biological warfare" for its high airborne infectivity, though it is less fatal than anthrax or plague.

Creating effective aerosols for most of these and other pathogens was a problem that only animal experiments could solve, with controlled temperature, humidity, and wind direction and velocity. Rosebury and Kabat recommended that this research be conducted by universities and they pointed out that public health benefits might also accrue from such investigations of airborne infection.

Second in importance to airborne agents were vectorborne diseases communicated by fleas and lice, mosquitoes, ticks, sand flies, or other vectors harboring parasites, whether bacteria, viruses, or protozoans. These possibilities, too, needed to be evaluated by experiments. The vectorborne diseases raised questions about which global areas might be attack targets. Mosquito-borne dengue fever, for example, characterized by relatively sudden onset (usually less than six days) with a slow convalescence, was identified with tropical climates. So, too, was yellow fever, also transmitted by mosquitoes and, although known in the Americas, of potential devastating consequence in Asia. Rosebury and Kabat suggested that some rickettsiae (the microorganisms in cells that can be transmitted by biting arthropods), like that for Q (query) fever, might be airborne. Tickborne bacteria causing spotted fever and Q fever were ranked high on infectivity, severity of symptoms, an extended convalescence, and possibly a high case mortality rate.

The two scientists more briefly discussed potential biological agents for causing animal and crop diseases. Among diseases that affect animals, anthrax, glanders, Rift Valley fever virus, and equine encephalitis could seriously affect both animals and humans. Foot-and-mouth disease, they noted, is highly infective for cattle, sheep, pigs, and other animals, but not for humans. Several viruses affect chickens and pigs. Rinderpest virus, a disease of cattle and buffalo, was noted for its severity; Canadian biological warfare scientists would later argue its threat to North American cattle and strongly recommend the development of a new vaccine. Bovine pleuropneumonia (*Mycoplasma mycoides mycoides*), eradicated from the Western hemisphere by 1895 and a danger to Asia, completed their list.

Rosebury and Kabat presented the possibilities of dozens of crop destroyers—bacteria, viruses, and fungi (most of them plant- and crop-specific), as well as insects like the boll weevil and the corn borer that might be launched against crops. Experiments, they again advised, would be needed to test the effectiveness of animal and plant pathogens and what technical defenses were possible. Plant pathogens proved the least favored among biological weapons options. Once released, they might prove hard to check or eradicate.

Military Defenses

Defending soldiers against possible biological weapons attacks was part of the mandate of all biological warfare programs. Troops needed to be protected against infectious disease agents that the enemy might use, as well as against the pathogens used by their own side. Following the chemical weapons model, a simple, well-fitted mask could filter out pathogenic aerosols. In the event of an attack, soldiers could don their masks quickly, but they would somehow have to be alerted that they were under attack by an invisible aerosol, which would be difficult or impossible under battlefield conditions. For some but not all agents, prophylactic or therapeutic drugs, tested for use after exposure, could help protect the soldier. Vaccinations or antisera promised round-the-clock protection, but not for most of the agents Rosebury and Kabat were ranking highly; furthermore, a large dose of an agent could overcome such protection. In addition, some biological agents might, as the authors noted, be combined or changed in laboratory experiments to enhance virulence. If the enemy chose an unknown weapon that intelligence sources had failed to reveal, that surprise factor could overwhelm all defenses.

No biological or chemical weapons were used during the war on any front. In fact, World War II proved a turning point in protecting soldiers in battle from disease and infection through technological innovation. Vaccines, sulfa drugs, the synthetic antimalarial drug atabrin and other derivatives, DDT (which protected against malaria and typhus), better techniques for the use of blood plasma and whole blood, and, later, the mass production of penicillin all contributed to better survival rates.

Civil Defense

Rosebury and Kabat did not flinch from the contradiction between biological weapons and the aims of public health. In public health, they observed, the main goal is to discover as much as possible about all the factors that spread disease, in order to break the causal chain. In contrast, to develop disease as a weapon reverses this goal of public health. As Rosebury and Kabat described, "For the student of bacterial warfare the emphasis shifts fundamentally. He is concerned with the weak links only in order to strengthen them, or to discard the whole chain if they cannot be strengthened. He is much more interested in the inception of mass infections than he is in the details of their perpetuation, except, of course, that for purposes of defense against bacterial warfare, or for control of mass infections once they have begun, he must fall back on knowledge of the whole epidemiologic chain."[35]

It followed logically that protection of civilians against biological weapons would be centered on public health reinforcement. Early detection was essential for the containment of a deliberate epidemic. Physicians, especially those in important industrial areas, had to be well informed. They needed to be alert to the use of potential agents, know how to diagnose the diseases caused by select agents, and be suspicious of unusual symptoms or circumstances in any disease outbreak.

In addition, Rosebury and Kabat proposed three levels of civil preparedness, which were never fully implemented but bear comparison with current American domestic preparedness plans. The first level was emergency response, which would require amplification or mobilization of existing public and private health service facilities. In support of this level, every large military establishment and every large industrial area should maintain mobile "medical bacterial warfare units" that would cooperate with local health, police, and civilian defense agencies and have the power to enforce quarantines, evacuate unaffected groups from dangerous areas, and be involved in distributing face masks and making sure that the appropriate vaccines and prophylactic and therapeutic drugs were available. These units would also be charged with informing the public about individual sanitary and other precautions once the agent was identified. Air raid and fire wardens could serve as decontamination and agent identification squads.

As a second level of preparedness, they recommended permanent or semipermanent measures for reducing the risks of respiratory diseases and of water and foodborne illness in workplaces and barracks. The methods they envisioned were those in use at the time: antibacterial sprays, ultraviolet lights, and chemical disinfection and sterilization.

As a third level of preparedness, Rosebury and Kabat addressed the opportunistic aspects of epidemics, advising broad public health measures to address social crowding, poor sanitation, and lack of vaccination and immunization in slums, already the breeding grounds of epidemic disease. "It would seem foolhardy in the extreme," they wrote, "to suggest that the possible consequences of a bacterial attack in such areas of congestion can be dismissed lightly."[36] Emergency measures would suit the military and industry, but the civilian population would also need long-term protective policies to reduce the consequences of epidemics in general.

This advice was given in the context of the great resurgence in public health that marked the administration of President Franklin Roosevelt.[37] New York City, where Rosebury and Kabat were located, had seen a tremendous growth in public health programs, for instance, in community health, child health, mosquito eradication, family nutrition, and dental care, and in new hospitals, local clinics, and laboratories. When advising emergency reinforcement of public health, the two scientists were referring to a strong infrastructure already in place, to which some but not great additions could be made. In the next fifty years, the American public health infrastructure was largely replaced by central hospitals, high-technology medicine, and curative care oriented to the individual patient as consumer.

After the war, Rosebury returned to medical science, writing about aerobiology and infectious diseases while also warning of the dangers of biological weapons. His 1949 book *Peace or Pestilence?* remains a classic that outlines the risks of biological weapons projects and asks that the public make itself aware of the gravity of the policy issues.[38] Kabat continued as a consultant to the Chemical Corps until 1947. He became a distinguished immunologist, receiving the National Medal of Science from President George H. W. Bush in 1991.

As the next chapter recounts, the tension between wartime public health and biological weapons or "public health in reverse" dominated British government discussion in the years just preceding World War II. As the war became a reality, public health advocates went one way, promoting general strengthening of the British healthcare system. At the same time, other civilian biologists of high caliber took a distinctly different path and joined with the military to develop biological weapons. The British program was not the world's first venture in this area. France and then Japan preceded it. Rather, the UK effort was the world's first technical and organizational success, the innovation that moved biological weapons off the drawing boards and into the reality of munitions, industrial production, and simulated air strikes.

CHAPTER 2

The United Kingdom and Biological Warfare

The Remorseless Advance of Military Science

In 1940, the British government embarked on a biological warfare program that began serendipitously and yet established an enduring organizational model of laboratory research and field tests of munitions. The UK initiative pioneered the way for the American effort, which then greatly surpassed it in size but not in ingenuity. The British program presents an example of "inadvertent escalation" under circumstances of war, with high officials initially unaware that research on a new type of weapon had been authorized.[1] British dread of Germany's destructive power motivated the decision to prepare retaliatory biological warfare capacity. Earlier suspicion of Germany's capacity and intent had already aroused British concerns about the threat of biological weapons.

Spies and Sabotage

German attempts to infect the Allies' pack animals with glanders and anthrax in World War I were minor exploits compared to the implications of the 1934 revelations of British journalist Henry Wickham Steed. According to secret documents Steed said he had received, German agents had investigated the potential impact of aerial attacks of chemical and biological weapons on London and Paris.[2]

The German use of airplanes and airships (especially the zeppelin) in World War I had caused dread among the British population. These German bombings killed 1,400 British civilians, a small number compared to British battlefield losses but a terrifying demonstration that the English Channel no longer served as a reliable protective barrier against traditional enemies on the continent. The distinction between combatant and noncombatant, already blurred by industrial-age war, could dissolve in air war, where the inaccurate bombing of military targets and the targeting of cities put civilians in double jeopardy. Steed's revelations, suggesting Germany's possible combination of airplanes and deadly aerosols, raised concern throughout Europe.

Wickham Steed began his article by translating an alleged July 1932 memo on strategic bombing from the commander of a secret German department, Air Gas Defense (Luft-Gas-Verteidigung): "As you were informed some time ago, the gigantic French fortifications on our Western frontier make attack by infantry seem quite, and artillery attack almost, hopeless. Consequently there remains only the most intensive development and extension of the air weapon, in order that air warfare may be waged effectively and ruthlessly against important military and industrial centres and, above all, also against the civilian population of large cities."[3]

Other German documents quoted by Steed outlined plans for spraying a usually harmless red bacteria, *Bacillus prodigiosus,* at entrances to the Paris Metro and the London Underground. This use of a simulant, the pathogen now commonly known as *Serratia marcescens,* made sense as a testing device. As presented, the idea was to track how air currents might spread toxic gas or pathogens through the entire subway system over, for example, a six-hour period. The premise was that bacteria or chemicals from a low-flying aircraft would be sucked into the ventilation systems of underground transportation. The subway systems in Berlin and Hamburg had already been tested, the documents claimed, in the event of an enemy attack by air. The documents also presented details about specific August 18, 1933, tests outside eight Paris Metro stops.

Not all experts were convinced of the authenticity of Steed's information. The documents contained internal inconsistencies and scientific notations that were not in German or even recognizably scientific. One critic pointed out that, after six hours, the stiff winds in Paris on August 18 would have blown the dispersed bacteria as far as the Belgian frontier.[4] Steed defended the authenticity of the papers he had received, on the basis of his sources, which he kept anonymous.

The clumsiness of the alleged German tests took nothing from their alarming intent. In public, British and French governments played down the significance of the Wickham Steed documents. Behind closed doors, British intelligence added Steed's information to another report received around the same time. Its spies had reported that an ongoing Berlin course for advanced military specialists in gas weapons included material on bacteriological warfare.[5] According to these sources, the teachers of the course insisted that Germany would only use biological agents in retaliation, thus adhering to the Geneva Protocol. This intelligence was deemed inconclusive, but the seeds of suspicion about a German biological weapons threat had been planted.

In this period, faulty intelligence on both sides led to misapprehensions of the enemy's chemical and biological warfare (CBW) strengths and intentions. At times intelligence was misjudged or ignored and government suppositions left much to fate. Given a reliable account from a German prisoner of war that the Third Reich had developed a nerve gas, tabun, the British government dismissed the information.[6] For their part, the Germans mistakenly believed the British chemical warfare program had also developed nerve gas, a belief that apparently deterred their own use of tabun, which they had manufactured by the ton. After the war, Reichsmarshall Hermann Goering told his captors that the Nazi command had resisted using nerve gas during the D-Day invasion for fear that an Allied retaliation in kind would kill the precious pack animals on which the German forces, low on gasoline for trucks and jeeps, were heavily reliant.[7]

In the buildup to World War II, the British and their Allies continued to receive intermittent intelligence reports about German biological weapons production, all of them suggestive and none, as it turned out, accurate. Adolf Hitler's personal aversion to biological weapons was a major restraint. German microbiologists, who speculated that the British might spray anticrop diseases or airdrop vials of lethal bacteria, were limited to small aerosol experiments. In addition to Hitler's objections, during the Third Reich the great legacy of German biology degenerated to eugenics and racist theories that led to the death-camp experiments and murders of Jews and Gypsies. Until the end of 1944, when it was certain that the Germans had no biological warfare capacity, the British justified their program's retaliatory goals with "looking-glass" presumptions that whatever their biological weapons program achieved, the Germans must also have accomplished.

The Man of Secrets

The British might have confined themselves to purely defensive measures against biological warfare—public health reinforcement and, for the military, vaccines, drugs, and face masks—had it not been for Maurice Hankey, longtime secretary to the British Cabinet and the most influential civil servant in the interwar years.[8]

Hankey's 1974 biography *Man of Secrets* makes no reference to his activities on behalf of the program, which continued after the war,[9] and his collected personal papers do not reveal his crucial influence. After years of historians representing him as antagonistic to the biological weapon program, Hankey was eventually acknowledged as its founding father by UK biological weapons scientists.[10] Newly declassified documents, reported on by British historian Brian Balmer, also document the centrality of Hankey's hidden role.[11]

Throughout his career, Hankey was a behind-the-scenes influence on British defense policy. During World War I, he was Secretary to the War Cabinet. In 1918, he became Secretary to the British Cabinet, a position he invented and then held for twenty years, while simultaneously acting as Secretary of the Committee of Imperial Defence. Hankey organized Cabinet and Defence agendas and consulted closely with Prime Minister Neville Chamberlain in the years leading up to World War II. He served on numerous select committees and created others that he often chaired or otherwise controlled. Officially retired in 1938, Hankey returned to government to aid the mobilization of the United Kingdom for the coming conflict with Germany. In 1939, he became Minister without Portfolio in the new War Cabinet, whose organization he had planned in detail, and he supported the successful proposal for a Ministry of Home Security to prepare defenses against German air raids. He also worked closely with leaders of another new bureaucracy, the Ministry of Economic Warfare, which spawned Special Operations Executive, a clandestine agency responsible for sabotage operations and later a client of the biological weapons program.

Biological Warfare and Science Advisors

Before the Wickham Steed documents were made public, they were given to Hankey, who brought them to the attention of three eminent scientists, John Ledingham, director of the Lister Institute, Professor William Topley of the London School of Hygiene and Tropical Medicine, and Captain Stewart Douglas, deputy director of the National Institute for Medical Research. All three were ambivalent about the authenticity of the documents; they thought the tests, if they actually occurred, related to chemical and not biological agent dispersal. Furthermore, the three scientists were skeptical about biological weapons in general.[12] Unlike Auguste Trillat, whose work they may not have known, they believed that the shock from explosive devices would kill most bacteria and that sprays from airplanes would result in aerosols too diffuse to cause serious, large-scale outbreaks.

In the aftermath of the Wickham Steed controversy, Hankey met privately with physiologist Sir Edward Mellanby, secretary of the Medical Research Council, to ask whether the Council might address the biological warfare threat. Mellanby emphatically rejected the idea. He saw the Council's mission as the promotion of medical research for healing, not for military ends, and certainly not to prepare weapons, even for retaliation.

Rather than retreat, Hankey cultivated Mellanby as an advisor and encouraged his project to improve the nutritional status of army recruits. In 1936, Hankey became chair of the Subcommittee on Bacteriological Warfare, within the Imperial Defense Ministry. He recruited Mellanby as a member, and also Ledingham and Topley and other top medical scientists (Douglas had by this time passed away). The task of the committee was to report on "the practicality of biological weapons" and recommend "countermeasures which should be taken to deal with such an eventuality." Mellanby and his scientific colleagues began meeting with representatives from the armed forces and from the Chemical Defence Research Establishment, most of whose budget went to the chemical weapons research program at Porton Down, west of London on the relatively remote Wiltshire plain.

As late as 1938, the science advisors Hankey had assembled doubted the practicality of germ weapons and therefore the German biological weapons threat. Instead, the committee's main concern was with the potential casualties from German air raids on the British and, secondarily, with the possible diseases that might result from the injuries, disruptions, and hardships of war. The most influential committee members had firsthand experience with large-scale epidemics. They had served overseas—in Mesopotamia, Serbia, India, and China—under conditions of war and social disruption that had generated serious outbreaks.[13] They were also of the generation that had experienced the 1918 influenza epidemic, which, as it circled the globe killing millions, demonstrated the opportunistic nature of epidemics. The committee concentrated on what it perceived as future wartime civilian vulnerability in the United Kingdom: the public health needs brought on by wounds from aerial bombings, threats to water supplies, and possible mass epidemics among civilians debilitated by wartime food and housing deprivations.[14]

Writing about the effects of total war on Great Britain, historian Arthur Marwick describes 1938 as a period when "politicians were obsessed with the problem of civilian casualties."[15] This obsession became the driving force for the creation of the Emergency Hospital Service, which required cooperation between voluntary and local hospitals regionally and nationally. By 1941,

80 percent of all British hospitals were integrated into the scheme and they opened their doors for free care to a wide variety of patients in need, not just bomb casualties and veterans. The Emergency Hospital Service became an integral part of the postwar framework for nationalized health care.

Attuned to public health, Hankey's committee prepared a long report proposing a national network of emergency bacteriological laboratories to aid in better diagnosis and treatment of civilian casualties. The report was passed to the Medical Research Council, headed by Mellanby, which established a special subcommittee to evaluate it and appointed Topley, one of its authors, as committee chair. In July 1938, the Committee of Imperial Defence approved the proposal for a national laboratory, which Mellanby named the Emergency Public Health Laboratory Service. This network of laboratories endured after the war as the Public Health Laboratory Service, an important part of disease management in the United Kingdom.

Until this juncture, the work of Hankey's committee had been highly secret. Now the members appealed to Hankey for permission to advertise their public health objectives openly. The Committee of Imperial Defence responded by allowing only limited contacts with nongovernmental organizations involved with the committee's other project, the stockpiling of pharmaceuticals.

After September, 1939, when Britain and Germany were officially at war, Hankey pressed his committee of advisors to focus on the technical potentials of biological weapons. While its members admitted that sabotage could frighten and demoralize the public and that botulinum toxin could be especially dangerous, they still rejected the notion that biological weapons could cause a serious epidemic. They were unimpressed by the increasing but fragmentary intelligence reports of German biological weapons technology: bombs, glass vials, and sprays. When asked to review the practicality of different methods, they stuck to their skepticism about the threat of aerosols and put their faith in routine, efficient public health responses. They discounted the dangers of anthrax, tularemia, and other noncontagious diseases. In a November 1939 report, Mellanby and the committee empathically rejected the airplane as "a very clumsy and inefficient bacteriological instrument" incapable of "depositing an infective agent in the right place at the right time."

Nonetheless, Hankey persisted in questioning his science advising committee about scientific uncertainties regarding disease agents. Had the reactions of bacteria to explosives been tested? How might anthrax spores travel through the air? Finally, Mellanby asked Hankey to get permission from Prime Minister Chamberlain for a series of aerosol experiments at the National Institute for Medical Research. Chamberlain agreed, after Hankey represented the experiments as purely defensive, in no way intended for weapons development. These tests, never fully reported, could have been either defensive or offensive in their purpose. Mellanby's request proved to be a first step toward more comprehensive exploration of the potential of anthrax and other potential biological weapons agents.

Sir Frederick Banting

Meanwhile, in Canada, Frederick Banting, head of his own research center at the University of Toronto, had become convinced that Hitler and the German army, as well as the Italians and the Japanese, would wage biological warfare and that the British government must establish itself in this field, offensively and defensively.[16] Born in 1891 in rural Ontario, Banting was a medic during World War I, and, like others, he knew how unprepared the British and their allies had been to retaliate quickly against German chemical weapons. After being trained as a surgeon, Banting immediately switched to medical science, and his subsequent discovery of insulin saved millions of diabetics from death. In 1923, Banting shared a Nobel Prize with John James Macleod for this work.[17]

Banting, a believer in the future importance of air warfare, envisioned the potential of airplanes at altitudes of 35,000–40,000 feet for dispersing bacteria over large areas. Although he knew little about bacteriology, he reasoned that infection could be imposed in a variety of ways. Conjecturing the use of airplanes, he declared that "under the new conditions of bacteriological warfare, the disease producing bacteria may be scattered anywhere—and it must be remembered that as long as bacteria gain entrance to the body they produce disease."[18]

Banting argued enthusiastically for the development of British biological warfare capacity. Without explicitly advocating the first use of germ weapons, he proposed that having an enormous retaliatory strength would give the United Kingdom a decided advantage over Germany. "If the Germans use bacteria," he asserted, "the Allies should be in a position to retaliate one hundred fold without delay."[19]

Banting set forth his ideas about retaliation, defense, and the German biological warfare threat in a position paper for the Canadian Defence Ministry. He listed tetanus, rabies, gas gangrene, parrot fever, anthrax, botulinum toxin, and, against livestock, foot-and-mouth disease as possible destructive agents, and warned of insectborne diseases and assaults on water reservoirs. This paper was passed on to officials in the United Kingdom, but his perspective was rejected in favor of the public health approach, already developed by

Hankey's science advisors. Edward Mellanby had presented this public health approach during a visit to Ottawa in September 1938 and it was the reigning government position.

In 1939, immediately following the German invasion of Poland and the British-French declaration of war, Banting, then forty-seven, rejoined the Canadian army as a major and devoted himself entirely to the war effort. While still head of his research institute, Banting chaired two high-level committees for mobilizing Canadian science. One was on medical research and the other on aviation medicine. For the latter, Banting invented the first program to investigate the effects of high altitude on pilots.

In November 1939, Banting traveled to England to lend his authority to the war effort. He and his scientific colleague, toxicologist Israel Rabinowitch, visited the Porton Down chemical weapons facility in Wiltshire and found its program in full swing, although, hampered by the war, it lacked the space for large-scale field trials. Rabinowitch suggested that Canada's open spaces could be useful. The nearly one-thousand-square-mile Suffield range in the province of Alberta later became a center for chemical and biological field tests.

Banting's more pressing mission on this trip was to persuade other influential scientists of the necessity to start a biological warfare program. He was convinced that Germany had been experimenting with bacteriological weapons for years, that it would not hesitate to use them, and the United Kingdom had to be prepared to retaliate by strategic attack. Banting took his case from one elite biologist to another with no results. On Christmas Eve of 1939, he had tea with Edward Mellanby, whose apparent lack of enthusiasm added to Banting's sense of discouragement. Banting also approached Maurice Hankey.

At this time, Banting was arguing that civilians were integral to modern war: "In the past, war was confined for the most part to men in uniform, but with increased mechanization of armies and the introduction of air forces, there is an increased dependence on the home country, and eight to ten people working at home are now required to keep one man in the fighting line. This state of affairs alters the complexion of war. It really amounts to one nation fighting another nation. This being so, it is just as effective to kill or disable ten unarmed workers at home as to put a soldier out of action, and if this can be done with less risk, then it would be advantageous to employ any mode of warfare to accomplish this."[20] Having accepted this grim calculation, Banting put his mind to how a biological weapons program might be structured. In the memorandum he brought with him to London, he proposed a new organization.

This new organization would have three functions. The first would be to gather and evaluate all known data on the feasibility of making biological weapons agents. For those skeptical of the stability of pathogens, Banting pointed out that organisms for typhoid, whooping cough, and meningitis had been successfully frozen and dried, and, by his reckoning, the problem of stability could be solved for other microbes. This part of the program would evaluate treatments to counteract these agents.

The second function of the program would be weapons research. Like others, Banting was interested in poison bullets. In addition, he advised experiments with shells and bombs and with the creation of a "bacterial dust" for dissemination from airplanes. He foresaw the need for factory-scale production of bacteria and a need for safety standards to protect workers. He theorized (correctly, as it turned out) that methods could be devised to bypass the vectors needed to transmit dangerous diseases such as plague and yellow fever.

The third part of Banting's proposed biological warfare program would be a "Vigilance Committee," a kind of intelligence wing that would track suspicious outbreaks and check up on possible espionage activities, such as the sale and use of laboratory equipment or other machinery indicating agent production. This committee was also charged with giving the public simple warnings, regulations, and instructions in the event of an outbreak; it might also, he suggested, consider the preparation of sufficient gauze nose masks for rapid, universal distribution if necessary to contain the spread of disease. Banting, alert to the many ways an enemy might wage bacteriological war against the United Kingdom, even considered the possibility of bacteria being sent in envelopes through the mail, as indeed they were in 2001 in the US. He was especially concerned about water sources being poisoned. "Fixed ideas of natural epidemics must be altered in the face of artificial warfare epidemics," he warned.[21]

He ended his memo by offering, if the British government requested, trained Canadian bacteriologists, vast open spaces for testing, and the cooperation, through Canada's Department of National Defence, of artillery and air force units.

In February 1940 Banting left London convinced that his biological weapons proposal had been rejected. He later dismissed Hankey as "a superb example of the servile, all-important complacent superior ass that runs the British government."[22]

The "Fog of War"

Banting very much misjudged Hankey, who was impressed by Banting's argument for biological weapons research and its placement alongside the chemical program. Soon after Banting left England, Hankey asked his science advisory committee (reconstituted under the War Cabinet as the Biological Warfare Committee) to consider the Canadian scientist's biological warfare proposal. Although the committee as a whole remained unconvinced, Hankey ended the meeting by suggesting that experiments should at least be tried, preferably at Porton Down, the site of the established chemical weapons program. He held a private discussion with Mellanby and other more amenable committee members. Hankey then appointed John Ledingham and a subgroup to research the possibilities of germ weapons. Ledingham reported back to the Biological Warfare Committee that there were three likely uses of germ weapons. One was contaminated rifle bullets, which experiments showed worked. The second was the use of germ agents in sabotage, which the Germans had already tried in the previous war. The third was aerial dispersion, about which more needed to be known.

The committee resisted recommending any new government action. To them, poison bullets seemed too dangerous for workers to produce and for soldiers to use. Sabotage with microbes or poisons could work, but it hardly required an entire weapons program to kill a few adversaries with ricin or to infect horses with glanders or anthrax. And if bacteriological aerosols were feasible, the committee reasoned, they would require a considerable state enterprise to reach a dangerous scale of attack. The committee members could hardly imagine that Germany would divert its wartime resources to this effort. In this assessment they later proved accurate, for Germany, but not for their own government, Canada, and the United States.

On May 10, 1940, Winston Churchill replaced Neville Chamberlain as Prime Minister and also claimed the office of Minister of Defence. The headlines that should have gone to Churchill's appointment went instead to the news that Germany had invaded the Netherlands. After years of service Maurice Hankey was now out of government. He and Churchill had previously clashed over a number of issues (including Churchill's tendency to argue government policies in the press). Churchill, though, recognized Hankey's special organizational talents and wrote to Chamberlain suggesting that Hankey should become Chancellor of the Duchy of Lancaster—an appointment "in which he can continue the chairmanship of the many secret committees over which he presides." In agreement, Chamberlain wrote "V. good" on Churchill's memo and underlined the notation twice to indicate his enthusiasm.[23]

In July 1940 Hankey, who then held no government office, wrote in a memo, "I recently came to the conclusion that we ought to go a step further in the matter of bacteriological warfare so as to put ourselves in a position to retaliate if such abominable methods should be used against us."[24] France had fallen to Germany the month before, and there were rumors that its secret biological weapons program was under German control. Germany was out to destroy British air power as a prelude to possible invasion and was bombing London and other cities to draw out the Royal Air Force. The Germans in 1940 had "the most effective short-to-medium range bombing fleet in the world and, unlike the RAF, German air forces had solved many navigational and bomb aiming problems."[25] Alone and under violent assault, Britain was entering what Churchill called "its finest hour."

With German bombs falling, Hankey made the decision that the United Kingdom needed a biological weapons program and, as he often did, he acted on opportunity. The previous year, in early August 1939, authority over the Chemical Defence Experimental Station at Porton had passed from the War Office to the new Ministry of Supply. Without consulting with either the War Cabinet, which he judged far too busy to consider germ warfare, or Prime Minister Churchill, Hankey conferred privately with Sir Edward Mellanby and two other of his committee members about the problem of how to initiate a biological weapons program. He and Mellanby then went to the Minister of Supply, Herbert Morrison, to ask that biological weapons be developed at Porton Down, alongside the chemical program, which the Ministry oversaw. In this period of extreme national duress, the argument for new weapons made sense. Morrison agreed, and on August 18, 1940, he authorized a Biology Department to be created at Porton for biological weapons research and development. Hankey's rationale for sidestepping government committees, the War Cabinet, and the prime minister was that the biological weapons research was exploratory, like other experimental ventures undertaken within the chemical weapons program.[26]

On the advice of John Ledingham, Hankey recruited Paul Fildes, head of the Medical Research Council's Bacterial Chemistry Unit, to head the new secret department. A bacteriologist, Fildes took the position believing that his mission was to achieve high-volume offensive capacity available at short notice. As a civilian Fildes understood that he would answer only to Hankey, rather than to the military. In early October, after arrangements with Porton had been made, Hankey had yet to inform the cabinet or the prime minister. He then received an offhand inquiry from Special Operations about possible crop destruction research. Hankey took this opportunity to write Churchill that the "practicability" of biological weapons had to be investigated in case it was ever used against the United Kingdom, at which point the military should be in a position "to retaliate if the Cabinet should so decide." Churchill approved. He had already pondered the future of biological weapons, writing in 1925: "Blight to destroy crops, Anthrax to slay horses and cattle, Plague to poison not armies but whole districts—such are the lines along which military science is remorselessly advancing."[27]

Thereafter Hankey kept Churchill advised of Fildes's work, but Fildes's goal of offensive capacity (in the name of retaliation) still lacked formal approval from the War Cabinet. Finally, in January 1942, the cabinet informed Hankey of its approval, with ample warning that the authority for using such weapons rested exclusively with itself or the Defence Committee and that the program should be strictly secret. The next day a note was appended warning Hankey that the new program must not divert resources from the ongoing war effort.

To be quickly prepared for retaliation in kind, Fildes started with linseed cakes, which tests at Porton had shown cattle would eagerly consume. He then had five million made in a London soap factory and shipped to Porton, where local women, working in a small building nicknamed the "bun factory," injected them with anthrax spores. The project was first called Operation Vegetarian, then renamed Operation Aladdin. The cattle cakes, however, were never used. It was calculated that a dozen Lancaster bombers in a single raid could drop enough anthrax cakes to threaten cattle throughout most of the north German countryside. Had they been used, they might have caused massive animal deaths and subsequent human illness, in addition to reducing food supplies and damaging the leather industry.[28] Since anthrax spores can persist in the soil for years, the cattle cakes could also have caused future outbreaks.

Fildes's main goal was to invent what Hankey's committee failed to imagine: a lethal anthrax aerosol widely dispersed by bombs. At Porton Down, experts in meteorology and physics, in munitions and in large-scale field trials over land and sea greatly assisted in achieving this goal. Fildes, an expert in growing bacteria, soon had a team working on a biological weapons model that approximated the one for chemical weapons. He recruited David Henderson, who was working on the chemical side of Porton and had done work on bacterial aerosols. By focusing on anthrax, with its tough spores and high virulence, Fildes had circumvented the fragile bacteria obstacle. With Henderson, he could investigate aerosols using anthrax spores. By the end of 1940, Fildes and his team had started small-scale spray and bomb tests at Porton.

Biological Weapons Funding from Industrial Giants

In the summer of 1940, as Maurice Hankey was covertly initiating the Porton Down biological weapons program, Frederick Banting was taking advantage of an unusual opportunity that would shape the parallel Canadian effort and, indirectly, the US and UK programs. Following the fall of France, three wealthy Canadian captains of industry stepped forward with millions of dollars for war-related research—John David Eaton, founder of T. Eaton department stores; Sir Edward Beatty, president of the Canadian Pacific Railway; and Samuel Bronfman, the head of Seagrams.[29] Banting was one of three scientists they consulted. The other two were engineer C. J. MacKenzie, head of the National Research Council, and Otto Maass, a well-known chemist committed to chemical weapons production.

Together, the six assembled a highly secret committee of science and military experts (the War Technical and Scientific Development Committee) to make decisions about innovative defense projects. Banting was soon awarded $25,000 to conduct laboratory experiments on germ weapons. Convinced of the necessity of both defensive and offensive biological weapons preparations, he organized around him a core of Canadian biologists and physicians to work on a wide variety of pathogens and on the physiological effects of chemical weapons. Banting's own laboratory at the University of Toronto was involved, along with the prestigious Connaught Laboratory, which was affiliated with the University and had a large facility ("the Farm") north of the city.

Banting's first biological weapons field experiment was to simulate an aerial attack. In early October 1940, near Balsam Lake northeast of Toronto, he succeeded in spraying sawdust from a low-flying plane. Afterward he and his assistants combed the ground to measure the dispersal path. The idea was not as far-fetched as it appeared. A few weeks later, one of Banting's junior scientists succeeded in drying typhoid bacteria on sawdust, a first step toward the mass use of a biological weapon against humans.

On the basis of his airplane experiment, Banting immediately went to the Secretary of Defense in Ottawa to argue for permission to produce bacteria for mass dissemination. Late that night, overwrought with patriotic feeling, he ranted in his diary about killing "three or four million young huns—without mercy—without feeling.... Those Huns at home, those huns of Hitler—It is our job to kill them."[30]

Banting's ideas on biological warfare found favor with the Canadian Prime Minister, MacKenzie King, and on November 19, 1940, Banting received permission from Canada's Department of National Defence for bacteria production. Banting thought in terms of large-scale manufacture, a factory where a hundred tons of virulent organism could be produced with safety and speed. No such production plant was forthcoming in wartime Canada, but Banting's industrial-scale vision was integral to the US program and those that followed.

Through C. J. MacKenzie, word of Banting's secret enterprise traveled directly to Vannevar Bush, president of the Carnegie Institute and founder of the National Defense Research Committee. Bush, an advisor to President Franklin Roosevelt, set in motion the first American inquiries into potential biological weapons strategies. Future contacts between US and Canadian scientists working on biological weapons proved direct and collegial. This was in contrast to the British presumption that Canadian scientists were junior partners, mainly just supplying the wide open spaces for UK testing.

On February 21, 1941, en route to London and from there most likely to Porton Down, Banting was killed in an air crash in the Canadian Maritime Provinces. Banting's death caused mourning throughout Canada. A few days later, on February 24, Paul Fildes released his first report on successful anthrax aerosol tests and their possible biological warfare applications.

Biological Weapons Experiments and Testing

Experiments and tests were vital to moving biological weapons technology toward assimilation by the military. Although anthrax proved tough and lethal in small tests, Fildes needed more precision in reckoning its virulence and infectivity. By April 1942, his team had perfected a cloud chamber for bacteria. The method of exposure was to hold the snouts of laboratory animals to apertures and let the aerosol pass by. Cloud chamber experiments facilitated determination of the aerosol particle size most effective for causing infection through the lungs and for estimating lethal doses for laboratory animals. For anthrax, a goal was to extrapolate from animal tests the amount of biological munitions and the number of aircraft that would be required to carry out a strategic aerial attack. Inhalational anthrax was known as a dangerous disease with no reliable therapy; to test it on humans would be impossible.

Fildes was eager to test the open-air effects of aerosols from anthrax bombs over a large area, on tethered sheep. Although relatively remote, Porton Down was too close to human habitation to pursue such potentially dangerous experiments. For this Fildes submitted a proposal to the Porton Executive Committee, which was then in charge of giving his projects oversight. Fildes had learned how to answer to government and, though never completely reconciled to military control over his research, he realized that his biological weapons needed "potential users," especially the Royal Air Force. Fildes received permission, and that summer a contingent from Porton's chemical and biological programs traveled to Scotland and set up a base on the mainland across from Gruinard Island, where they would conduct historic bomb tests. Later during the war they would also conduct biological weapons tests at Penclawdd, on the coast of Wales.

It was in Fildes's favor that Prime Minister Churchill was open to unconventional weapons and already supported chemicals. Determined that the United Kingdom should have quick and massive capacity for chemical weapons retaliation, he pressed for increased production. The United Kingdom soon had twenty thousand tons of poison gas in storage, and 15 percent of its bomber force ostensibly at the ready to use it. Churchill also advertised this capability to the enemy. In an impassioned radio address on May 10, 1942, reacting to word that the Germans intended to use chemical weapons against the Soviet Union, now a Western ally, Churchill warned: "I wish now to make it plain that we shall treat the unprovoked use of poison gas against our Russian ally exactly as if it were used against ourselves, and if we are satisfied that this new outrage has been committed by Hitler we will use our great and growing air superiority in the West to carry gas warfare on the largest possible scale far and wide against military objectives in Germany. It is thus for Hitler to choose whether he wishes to add this additional horror to aerial warfare."[31] This same fierce declaration inspired the American scientists Theodore Rosebury and Elkin Kabat to propose a retaliatory capacity for biological weapons in their 1942 report.

In July of 1942, as the trials at Gruinard Island were commencing, Churchill may have suspected that the Japanese, known to have chemical weapons, had developed biological weapons as well. At that time, the Chinese appealed to the West with accusations that the Japanese Imperial Army had spread plague against cities in eastern China. Although British colonial administrators dismissed the allegations, Churchill personally marked the report for circulation and discussion among his defense experts.

The First Field Tests

A formidable scientist from Porton, meteorologist Sir Oliver Graham Sutton, led the research team of fifty men to conduct the Gruinard trials. Henderson and other staff were in charge of the germ weapons, and Fildes was a frequent visitor. The team also included Chemical Defence personnel and military engineers, physicians, veterinarians, and munitions experts. They camped on the mainland, where rudimentary housing and a small laboratory were built. Gruinard was an uninhabited island about two kilometers long and one kilometer wide, covered with grass, heather, and bracken. About 155 sheep bought from local villagers were used in the tests.

The first trial at Gruinard, on July 15, used a modified thirty-pound bomb filled with three liters of anthrax suspension. Instead of risking the imprecision of an airplane drop, the bombs were suspended on gallows four feet above the ground and then detonated by remote control, producing a visible cloud that drifted downwind. After seven days, nearly all the sheep that had been tethered in the range of ninety to one hundred yards from the source were dead. Further tests showed that the anthrax cloud was lethal at up to four hundred yards. For a more realistic scenario, in September a Wellington bomber dropped an anthrax bomb from seven thousand feet; but the bomb landed in a peat bog and had no effect on the sheep. The team turned to another site, one previously used by Chemical Defence, at Penclawdd, in a beach area. There a bomb dropped from a plane at around five thousand feet missed its mark by only twenty feet. Sheep in the range of 120 to 320 yards from the explosion died soon after from anthrax. Trials continued at Gruinard throughout July and September 1943 and their results were reinforced by experiments with simulants at Porton.

The Gruinard tests suggested that, weight for weight, anthrax could be one hundred to one thousand times more potent than any chemical weapon of the time, to a distance of five hundred yards. Further, an efficient electrically fired four-pound bomb was developed; 106 of them could be packed into a cluster bomb derived from models used for incendiaries in World War I. Fildes had also defined the approximate lethal dose for sheep, an important standard for further research and production. These tests, only the first of many to follow, also began training soldiers through simulated biological warfare.

In August 1943, the tests at Gruinard were forced to a halt. The previous spring, after a heavy storm, dead sheep buried on the island near the coast apparently were dislodged and floated to the mainland, where they caused more than fifty deaths among livestock. The government told the alarmed villagers that a nearby Greek ship on maneuvers had tossed some infected carcasses overboard and compensated them for their lost animals. Wary of more accidents, Fildes hoped that future tests could be conducted in the open spaces of Canada and the United States.

Fildes's claim to the first modern biological weapon circulated through the various committees charged with national defense. The favorable comparison of anthrax to much heavier chemical agents meant that biological bombs might have a future. Fildes's success was taken as proof that the Germans must have developed a similar biological weapon, although there was no intelligence information to support this conclusion. Fildes himself firmly believed that Germany had an identical weapon and he continued to argue that the power of retaliation was the United Kingdom's best defense against the threat. His technical breakthrough, bombs that killed sheep, became the "looking-glass" justification for a full and ready offensive biological warfare arsenal.[32]

As he succeeded in his research, Fildes defended biological weapons with the humane argument that during World War I the Germans had used for chemical weapons, and which the noted British geneticist J. B. S. Haldane had also argued.[33] Those arguments referred to soldiers. Fildes saw his efforts as benefiting civilians as well. In response to a criticism of biological warfare as "unsoldierly" and eliciting "public revulsion," Fildes wrote: "Is it any more moral to kill Service men or civilians with HE (High Explosives) than with BW?...It seems clear to me that a substantial majority of the population would conclude that, if they had to put up with a war again, they would prefer to face the risks of attack by bacteria than bombardments by HE."[34] From Fildes's point of view, no moral or technical obstacles prevented the next logical step, the large-scale production of anthrax bacteria and thereafter the mass production of anthrax bombs. From a strategic perspective, the anthrax bomb opened up new potentials for destruction, uncoupled from retaliation.

Fildes was constantly in communication with the Canadians and Americans about the successes of the anthrax bomb, which he wanted mass-produced. Britain was deeply mortgaging itself to pay for the war, and he had been warned that biological warfare projects must not burden this effort. Canada had neither the skilled labor force nor the other industrial resources to spare. Fildes had taken the United Kingdom as far as it could go as with biological weapons. For industrial manufacture, his project needed the United States.

CHAPTER 3

The United States in World War II

Industrial Scale and Secrecy

The 1940 establishment of the Biology Department at Porton was an organizational innovation that pointed the United Kingdom toward a retaliatory biological weapons capacity of strategic proportions, a goal espoused by no other nation at that time. Producing five million anthrax cattle cakes was the program's first response, to be ready for retaliation should Germany attack with any kind of germ weapon. The prototype anthrax bomb tested at Gruinard and Penclawdd was the next, more serious effort. That seemingly small success opened the door to Frederick Banting's vision of biological weapons on an industrial scale, which the United States then pursued, using its ample resources to advantage. After the war, the US biological weapons program became larger still.

Although the Gruinard anthrax bomb tests spelled progress, serious technical obstacles stood in the way of biological weapons being integrated into Allied war plans. Large volumes of anthrax spores had to be produced to fill bombs; the spores had to be freed from unsporulated bacteria and other debris, with their virulence intact, and be dispersed in the air in the small particle sizes that would allow inhalation deep into the lungs to cause infection. Calculating dose responses for human victims was the toughest problem. Dose response had to be estimated from animal experiments in order to calculate the number of munitions required to mount an effective strategic attack, but the estimates were not necessarily accurate. The design of the agent fill and the explosive charge for various anthrax bomb models also posed persistent problems. Once the early bombs were developed and field tested, it was soon realized that they had to be manufactured in the hundreds of thousands in order to achieve strategic dimensions. The British at home were already supporting four chemical weapons plants and taking the risks of production and storage close to airfields, so the bombs could be readily available for loading. No similar, additional commitment could be made to the anthrax bomb.

In the summer of 1942, the US Chemical Warfare Service (CWS) was in touch with Paul Fildes and with Canadian biological weapons experts with its ideas for the US biological warfare program and the sites, facilities, and personnel it would need. In the autumn of 1942, while the anthrax bomb was still a work in progress, Fildes and David Henderson, his aerobiology expert and lead scientist, traveled to Canada and the United States in search of support. Maurice Hankey had paved the way for Fildes's trip with British Special Operations, which took special interest in biological weapons and Fildes's endeavors. The more successful his experiments were, in fact, the more his work was subject to control from different quarters of the defense establishment. This supervision took bureaucratic shape in multiple oversight committees that crosscut the Cabinet, the Ministry of Supply, and Ministry of Defence and included independent scientific advisors who did not always support Fildes's projects.[1] For this 1942 trip, he was required to emphasize the United Kingdom's defensive posture and its extreme reluctance to initiate the use of biological weapons, although retaliation could not be excluded.[2] By persuading Canada and the United States to support the bomb venture, he might circumvent government objections about its diverting resources from the UK war effort.

Arriving in Canada, Fildes and Henderson found scientists who were willing to help and had project ideas. On reaching Washington in November 1942, Fildes was disappointed that, despite the CWS's interest, the US government had made only preliminary inquiries into biological weapons, defensive or offensive.

US Science Advising on Biological Weapons

From the summer of 1941 to the time Fildes arrived, civilian science advisors dominated the secret discussion and planning for a US biological weapons initiative. The role of scientists in setting directions in military research had been institutionalized by the National Defense Research Committee (NDRC). President Roosevelt's science advisor, MIT engineer Vannevar Bush, had set up the committee for scientists to give advice on weapons technology. As the war in Europe intensified, biologists became drawn into plans for a biological weapons capacity.

In July 1941, the War Department convened a meeting with representatives of the Office of Scientific Research and Development, the Surgeon General's office, the Chemical Warfare Service, and Army intelligence (G-2) to brainstorm about the threat of biological weapons and the possible creation of a US program. This group recommended further study and, as a result, Secretary of War Henry Stimson requested a report from Frank Jewett, president of the National Academy of Sciences (NAS). Jewett passed the Stimson assignment to University of Wisconsin biologist Edwin Fred, who put together the secret War Bureau of Consultants (WBC) committee of twelve scientists, with liaison members from the Chemical Warfare Service, the Surgeon General, the US Public Health Service, the Department of Agriculture, and scientists from the US Army.

As the committee researched its report, the Japanese attack on Pearl Harbor brought the United States into the war. Particularly with reference to gas warfare, Stimson decided that the United States would not be bound by the terms of the Geneva Protocol, which might "through introduction of domestic, political and moral issues, impede our preparation, reduce our potential combat effectiveness and be considered, by our enemies, an indication of National weakness."[3]

Born in 1867, Stimson was an elder Republican statesman who had been secretary of war under President William Howard Taft and secretary of state under President Herbert Hoover. In 1932, after the Japanese invaded Manchuria, he framed the Stimson Doctrine, which declared that the United States would refuse recognition of situations or treaties that might threaten its interests or that were achieved by unprovoked aggression. In 1940, when he accepted Roosevelt's request to become secretary of war, he was read out of the Republican Party. He outlived Roosevelt and, under Truman, argued in favor of using the atomic bomb on Japan.

In February 1942, the completed NAS report was on Stimson's desk. Its authors concluded that biological weapons posed a serious threat to national security and that the United States should be moving forward with offensive and defensive measures. On April 29, 1942, Stimson sent the NAS report and his summary of it to President Roosevelt. The summary strongly emphasized the danger of biological weapons for human, plant, and animal life. Referring to biological warfare as "dirty business," Stimson argued for secrecy and vigor in pursuing defensive and offensive options. As Stimson saw it, appropriate defense relied on a thorough study of means of offense, and he urged that the United States be as prepared to retaliate with biological weapons as it would be for chemicals.[4] At a meeting two weeks later, Roosevelt verbally approved the committee's recommendations to explore biological weapons. From this point on, the question of whether Stimson kept the president informed about the growth of US biological weapons program remains open.

Roosevelt was almost surely as opposed to the first use of biological weapons as he was to the first use of chemical weapons, without being averse to having a wartime retaliatory capacity or to supporting the British. In the winter of 1940, the Americans began secretly supplying poison gas to the British.[5] On joining the war, the United States built two enormous chemical weapons facilities, one at the fifteen-thousand-acre Pine Bluff Arsenal in Arkansas, which at its most productive employed as many as ten thousand people, and the other at the twenty-thousand-acre Rocky Mountain Arsenal near Denver, which employed three thousand. Together the two produced millions of chemical bombs and artillery projectiles, as well as tens of thousands of tons of bulk Lewisite, mustard gas, and phosgene. In 1942, the US government also opened the Dugway Proving Ground, Utah, which encompassed more than a quarter million acres, and whose Great Peak was later designated for biological testing.

Roosevelt threatened retaliation with chemical weapons against Japan in 1942 and against Germany in 1943.[6] Unlike Churchill, though, he was in principle against using them: "I have been loath to believe that any nation, even our present enemies, could or would be willing to loose upon mankind such terrible and inhumane weapons."[7]

Admiral William Leahy, Roosevelt's close friend and his chief of staff in the war years, was adamantly opposed to biological weapons, as well as chemical and nuclear ones, and considered total war doctrine a modern form of barbarism. In his autobiography he writes how he expressed this opinion to Roosevelt in an informal July 1944 discussion with other advisors: "Mr. President, this (using germs and poisons) would violate every Christian ethic I have ever heard of and all of the known laws of war. It would be an attack on the noncombatant population of the enemy. The reaction can be foretold—if we use it, the enemy will use it."[8]

Biological weapons clearly had the potential to be controversial. In response to the WBC committee report, the War Department General Staff recommended that a civilian agency undertake program planning, and do so secretly, as a security measure and to prevent public alarm.[9] The Federal Security Agency, a 1939 innovation that included the Public Health Service, the Food and Drug Administration, and the Children's Bureau, was selected as a cover for a new defense research and development organization, the War Research Service (WRS), which was under Army and Navy authority.

In August 1942, Stimson appointed George Merck, president of the family pharmaceutical concern that bore his name, to head the WRS.[10] One WRS branch was designated Consultants in Research Policy, with two science advisory divisions, the Liaison Group and the ABC Committee (a scrambled acronym), formed with the help of the National Academy of Sciences. The second branch of the WRS was called Advisors on General Policy, including Jewett at the NAS, Vannevar Bush at the Office of Scientific Research and Development, and Harvard University president and chemist James Conant at the NDRC. The third branch of the WRS included Dr. Fred, in charge of research and development; Lt. Col. Arvo Thompson, a veterinarian who would later be involved in the Japanese immunity bargain, on technical aid; and the novelist J. P. Marquand in charge of the Information and Intelligence Division, which had ties to Army G-2, the Office of Strategic Services (OSS), Office of Naval Intelligence, and the Federal Bureau of Investigation. Thus began US mobilization of military and private resources for defensive and offensive projects.

The first project of the WRS was to provide instruction for military medical and security officers on detection and defense measures and to require status reports of their plans and operations. Its next was to establish twenty-five laboratory research contracts with US universities and institutes. Parallel to this research, pharmaceutical companies (including Merck, Pfizer, and Squibb) were closely involved in government defensive efforts to produce vaccines, drugs, sera, and, starting in 1943, the synthesis and mass production of penicillin.[11]

Paul Fildes's November 1942 visit to Washington came at a time when the CWS was positioned to assume more responsibility for biological weapons research and development. The CWS greeted the anthrax bomb project with favor, as did civilian scientists recruited to work on it.

Canadian Partners

Through Frederick Banting's influence, the links between biological weapons research and the Canadian Army's Directorate of Chemical Warfare and Smoke were established as early as 1940. Banting and chemistry professor Otto Maass guided the scientific projects investigating both kinds of weapons, and Canada's National Research Council assisted in coordinating these projects with military objectives. Enjoying considerable autonomy, Canadian scientists were sometimes able to plan chemical and biological projects with

British and American officials, and then bring their own government on board with approval and funding.

Following Banting's death in February 1941, the Canadian impetus to explore biological weapons faltered.[12] Then, in November of that year, with impetus from the Canadian Medical Research Committee that Banting had chaired, a group of scientists met to reconsider what research might be done for the war effort. They discussed plague and fleas and insect vectors in general, plus typhoid, botulinum toxin, and rinderpest, a cattle disease of Asia and Africa that they feared could wipe out North American herds if used in an attack.

The US entry into the war in December resuscitated Banting's agenda. Shortly after, an official from the US Chemical Warfare Service arrived in Toronto to review Banting's files and discovered anew his 1939 proposal for a biological weapons program and other memos and reports. Canadian scientists were poised to help. The new C-1 Committee under the Directorate of Chemical Warfare was authorized to organize projects. With the British anthrax bomb objective in mind, its director, E. G. D. Murray, selected Grosse Isle, a remote outpost in the Saint Lawrence River, where abandoned quarantine buildings might be refurbished for mass-producing anthrax spores. The Canadians invested $40,000 for reconstruction and equipment, in hopes that the United States would also contribute.

One of Stimson's first agreements with the Canadians was to fund the Grosse Isle station, not for anthrax but for a rinderpest vaccine study. The ABC Committee, fearing the devastation the disease might cause US livestock, had given rinderpest highest priority. Stimson then secretly bent the rules to facilitate covert US support of this and future Canadian projects related to biological weapons. Instead of pursuing a diplomatic agreement, which would have involved Congress, the State Department, and other agencies, Stimson arranged a contract whereby the US government would purchase certain secret reports at $15,000 each from a Canadian government agency. These reports would be sent to the head of the US Chemical Warfare Service, and from there to George Merck and the WRS.

This contract arrangement was soon used to fund US support of anthrax spore production at Grosse Isle. To secure funds for the anthrax project, Merck, Fred, Murray, and Canadian biologist R. W. Reed met with Gen. William Porter, the head of the CWS, and his top aides to persuade them that Canadian anthrax production warranted support. Convinced, Porter then paid $50,000 in return for secret reports on anthrax. Otto Maass acquired another $25,000 from the Canadian Ministry of Defence by assuring them that George Merck was backing the project.

The Grosse Isle station had a difficult, ultimately unsuccessful history. Isolated in northeastern Canada, secret, and hard to access in the winter, it was troubled by poor morale, breakdowns in safety standards, and unforeseen anthrax spore contamination. These setbacks were compounded by fears that anthrax-infected flies from the laboratory were contaminating food in the mess hall. Canada produced enough anthrax spores for bomb tests at Suffield and to share with Americans, but not more. By August 1944, the US program, led by Ira Baldwin, a yeast fermentation expert from the University of Wisconsin, was far ahead in volume production, and Grosse Isle was shut down.

The Canadians continued to serve the British need for agent development and testing. By March 1941, a British official from Porton was in charge of Experimental Station Suffield in Alberta, the proving ground known today as DRES (Defence Research Establishment Suffield). Conflicts later emerged over whether British or Canadian scientists were in charge of Suffield field experiments. The British at times failed to notify Canadian officials of their projects, including some involving lethal pathogens.

The British and Americans had better relations. During the war, David Henderson spent a good deal of time at Camp Detrick in Frederick, Maryland, where the American biological weapons program was established in 1943. The British bacteriologist Lord Trevor Stamp circulated between Camp Detrick and Suffield, where he conducted open-air tests, most likely of anthrax.[13] After the war, Henderson and Stamp each received the American Medal of Freedom for their work on biological weapons.

Preliminary Testing

The year 1943 marked a rapid acceleration in the US biological weapons program, as it did in America's general wartime mobilization. The Special Projects Division (SPD) was set up within the Chemical Warfare Service and Camp Detrick, which had been an unused airbase, became its center. SPD was highly secret, and its technical director and other scientific staff answered directly to Washington or to Edgewood Arsenal CWS officers, thus circumventing the commanding officer at Camp Detrick.[14]

From the very beginning, responsible officers in the United States, Canada, and the United Kingdom maintained the strictest secrecy about the work being done on biological warfare. They took stringent security measures not only to prevent the enemy from obtaining information, but also to keep the public and the rest of armed forces in the dark. Camp Detrick and other installations were set up as classified exempt stations, and elaborate precautions were taken to conceal their purposes.[15]

Granite Peak, at Dugway Proving Ground, was an SPD installation for biological field tests, including some with small quantities of anthrax to test its endurance in soil. Horn Island, off the Mississippi coast near Pascagoula, was an additional field post, used mainly for insect studies but also for aerosol dispersal tests of the anthrax simulant *Bacillus globigii* (BG).

The Americans built on the British aerosol chamber experiments to simulate dispersal and exposure data. The British had used the chamber to test agents for anthrax and plague on laboratory animals.[16] US scientists modified and enlarged the Porton model, so that they could put up to a hundred mice or fifty guinea pigs, rats, or hamsters inside it. In experiments with "hot" agents, the contaminated coats of the exposed animals were a risk to those who had to handle them afterward. Still, US scientists thought their model better simulated field exposure. Over time they invented other models that used airlocks and "animal storage boxes" attached to the chambers.[17] With cloud chamber studies, researchers could begin determining which inhaled agents were most infective, most lethal, and most stable.

As in the United Kingdom, though, the momentum was toward outdoor field trials, and the development and testing of an effective bomb was given high priority in the US program. The Porton bomb the Americans worked on was the Type F, a four-pound high-explosive model designed for chemicals but modified for liquid biological fill. It was designed so that 106 of these bombs would fit the British #14 cluster adaptor, which, as it hit ground, permitted the bombs to go off in succession. The total capacity of the Type F bomb was 400 cc; the recommended agent filling was 320 cc, which was 17.6 percent of the total bomb weight.

After testing the Type F bomb at Camp Detrick, engineers and technicians modified it to make a bomb that could be more easily mass-produced. After many minor changes were made, including in the fuse and the interior coating, the bomb was renamed the SPD Mark I.

The British claimed, or so the Americans understood, that one cluster bomb of anthrax could produce a cloud effective against humans a mile away. But making the bomb disseminate enough anthrax spores to cause significant fatalities, as best as could be estimated, remained a problem. Tests with anthrax simulants showed that neither the Type F nor the Mark I made efficient use of the agent. The emitted "clouds" were small; only 22 cc of the 320 cc of the agent went into the aerosol. Yet, under great time pressure, the scientists in charge decided that the Mark I was good enough for mass production. In May 1944, a pilot plant at Camp Detrick produced an initial run of five thousand anthrax-filled bombs.[18] Detrick was also able to fulfill an order of two thousand Type F bombs for the Canadians to test at Suffield.

Industrial Bomb Production and Churchill's Order

The pressure the Americans were feeling to mass produce anthrax bombs was coming from the highest level of British government. Yet within the committee structure that oversaw Fildes's Biology Department, some advisors strongly objected to his anthrax bomb project. Two of these, Victor Rothschild and Richard Peck, were members of the Biological Warfare Committee (BWC), which Maurice Hankey had organized but had left.

Air Marshal Richard Peck was a critic of strategic bombing in general and of what he felt was unfounded optimism about air warfare. In 1939 Peck reckoned that the invention of faster fighter planes, plus the development of radar, meant that enemy bombers could be defeated as they attacked. During the Battle of Britain, in July–September, 1940, when the Luftwaffe attacked Britain, the Royal Air Force's Hurricane and Spitfire fighters saved the day, due in part to Peck's influence.[19] By the same token, as Peck warned, Allied bombers were vulnerable to Axis fighter planes, anti-aircraft artillery, and radar. As the British soon found out, daylight raids on Germany meant heavy losses. Peck had also chaired the Porton Experiments Committee, the first one set up to oversee Fildes's work.

Victor Rothschild was a BWC member trained in zoology. A member of the prestigious banking family, he was also posted to British military intelligence. In 1939, he had met Frederick Banting and had not encouraged his biological weapons plans.

In December 1943 Paul Fildes announced to the BWC that the US government was preparing to meet the UK demand for mass-producing the anthrax bomb. This came as a surprise to committee members. Afterward, Peck, who thought Fildes was "dangerous," reacted by questioning his sense of responsibility and his judgment of "the consequences of loosing these inventions upon mankind."[20] Rothschild immediately complained to the war cabinet and set in motion an inquiry as to the authority behind this important decision to move to bomb production for aerial attack. A review of past files by the committee secretary revealed little except that, in the autumn of 1942, the war cabinet had authorized Hankey to proceed with the anthrax cattle cake project. At an impasse, the BWC referred the bomb project decision to Prime Minister Churchill.

Churchill passed the evaluation of the anthrax-bomb production proposal to his science advisor and confidant, Lord Cherwell, the architect of the strategic bombing campaign just commencing against continental Europe. Cherwell—the former Frederick Lindemann, a physics professor who had been head of the Clarendon Laboratory at Oxford University—was almost the only science advisor Churchill trusted. Cherwell had scant communication with other scientists once Churchill gave him his own laboratory and the two began to share imaginative military weapons schemes. One of these was the production of tanks fitted with bright lights to dazzle enemy gunners in night battles. Another was "snow torpedoes" for northern combat. Still another was warships made of reinforced ice. Cherwell earned the nickname "Lord Barrage of Balloons" after he arranged to have tethered balloons flown over London to discourage German bombers from approaching.[21]

Stressing the efficiency of biological bombs containing anthrax (the agent was code-named "N") over chemical ones, Fildes personally presented his case to Lord Cherwell. Cherwell was persuaded. In February he wrote Churchill, "We have developed what we believe to be effective means of storing and scattering N spores in 4 lb. bombs which go into ordinary incendiary containers. Half a dozen Lancasters could apparently carry enough, if spread evenly, to kill anyone found within a square mile and to render it uninhabitable thereafter."[22] With a reference to tube alloy, meaning the ongoing US atomic bomb project, Cherwell made what is perhaps the first comparison of biological weapons with nuclear ones: "This appears to be a weapon of appalling potentiality; almost nothing more formidable, because infinitely easier to make, than tube alloy. It seems most urgent to explore and even prepare the counter-measures, if any there be, but in the meantime it seems to me we cannot afford not to have N bombs in our armoury."[23]

Churchill secretly discussed this memo with his top military advisors, who apparently agreed with Cherwell. On March 8, Churchill ordered the chair of the BWC, Ernest Brown, to place an order for half a million anthrax bombs from the United States. Churchill, in the darkest days of the war, noted that this would be a "first installment."[24]

The Vigo Plant

The major US biological weapons enterprise of the war was the refurbishing of an ordnance plant in Vigo County, Indiana, for anthrax bomb production.[25] The US government acquired the six-thousand-acre plant, six miles from downtown Terre Haute, in 1942. The factory quickly constructed there to make explosives and detonators soon became outmoded and was shut down after three months. In November 1943 the Ordnance Department declared the Vigo plant surplus and turned it back to its builders, the Army Corps of Engineers, for disposal. Parts of the site were then leased to industry for storage.

In May 1944, the CWS acquired the Vigo plant. It was designated an SPD post for the manufacture of biological agents and vaccines, the filling and loading of biological munitions, and the breeding of laboratory animals. In addition to the plant, the post had its own rail line, 4.4 acres for detonating and burning explosives, three 14,000-gallon underground storage tanks, forty-six magazines, and a large warehouse, plus barracks for more than a thousand military personnel. At an early July meeting of representatives from CWS, the Corps of Engineers, the Surgeon General, and a private contractor, the group proposed a production and storage agenda suited to the British order for 500,000 bombs and more. More than that, the plan was to produce as many as a million bombs a month. The Vigo plant required an initial $10 million investment for startup and cost nearly $30 million by the war's end. Before Vigo could go into production, SPD scientists and technicians had to test high-volume anthrax spore production and assembly-line bomb production (including detonators and fuses) and bomb filling.

While the workers at Vigo were limited to trial runs with the anthrax simulant (again, *B. globigii*), two pilot plants were created in unused hangars at Camp Detrick. The first was a small-scale laboratory that produced seed cultures of anthrax; the second was designed to achieve volume production. Pilot Plant 2 was for making large volumes of slurry, which were then pumped into tanks in nearby buildings where the slurry was treated for sporulation and then pumped to another building for final processing and storage in two 6,000-gallon tanks. In the final phase, the spore material was pumped to three 1,000-gallon storage tanks in still another building, where it was funneled to special bomb-filling machines. The bombs were monitored for leakage and checked for the viability and virulence of the anthrax fill. The Vigo plant was designed to replicate this process on a grand scale.

Growing small quantities of anthrax in a laboratory and causing it to sporulate was relatively easy. Nor was maintaining its virulence a problem. The British had supplied the Americans with anthrax strains whose virulence had been enhanced by passing them through successive monkey hosts (the label M-36, for example, referred to serial passage through thirty-six monkeys). Scientists at Detrick were able to enhance *B. anthracis* virulence by 300 percent. Throughout the summer of 1944 and into the spring of 1945, the main difficulty proved to be producing the tons of anthrax spores needed for the British order and for

US stockpiles. Vigo was equipped with ten 20,000-gallon fermenters to get the job done. Among other difficulties, the anthrax bacteria could not be grown without adequate aeration, which the large vats failed to provide. In addition, the Detrick batches of anthrax were often contaminated by other bacteria, and the mechanical process of reducing an anthrax slurry to "mud" and preparing it for insertion in bombs was dangerous to workers.[26]

The live-spore Pasteur type vaccine protected livestock against anthrax but was considered too dangerous for people. Before work was started at Camp Detrick, the WRS commissioned a private firm to investigate and test immunization. The effort failed, although some progress was made testing penicillin as an antidote. At Camp Detrick, in November 1943, scientists in what was called Division B were directed to "to develop immunity against anthrax through analysis and investigation of various antigens and through serological studies."[27] Preliminary research on anthrax vaccine and antigens was begun, and the Detrick scientists gained rudimentary understanding of the deadly toxins made by anthrax bacteria. But there was no breakthrough with a human anthrax vaccine. Meanwhile, the pilot work on anthrax at Camp Detrick plodded on, leaving Vigo bomb production stalled.

Personnel Problems

Ira Baldwin, the fermentation specialist from University of Wisconsin, headed scientific research at Camp Detrick. He had selected the Vigo site and was subsequently in charge of refurbishing it.[28] As Baldwin later recounted, recruiting enough of the right people for any of the SPD posts had problems, not as severe as those at Grosse Isle but of the same type.[29] One was the sense of isolation that personnel experienced because of both the rural setting and enforced secrecy. Once recruited, SPD personnel were subject to strict security regulations, and not only the public but also the rest of the military knew nothing about its activities. Its personnel were also exempted from work on the Manhattan Project for an atomic bomb, to keep the two top-secret programs separate. Recruits were exempted from overseas duty, for fear they might be captured by the enemy. Even their vaccinations were considered a possible tip-off to the enemy about US research. Top-grade workers were not always available. At Camp Detrick, enlisted men who had already served overseas were frequently brought into the project; according to reports, many suffered from "psychoneurosis" and were soon discharged.

At Vigo, Navy and Army personnel came into conflict. The Navy personnel were more highly trained and had more lenient furloughs and discipline and they had higher expectations about housing; for instance, they wanted their barracks painted. The Navy later agreed to reduce its furloughs and leaves to Army levels and gave up the demand about painting the barracks. The Army apparently never provided the number of enlisted men it originally promised, and, by one report, between a third and a half of the Army recruits were unfit, sick, or fit for discharge.[30]

In July 1944, in response to an inquiry from Prime Minister Churchill, the UK Joint Planning staff estimated that the N-bomb was capable of making a difference in the war, but the Americans were behind in their production schedule. At this time, when the Allies were intensely bombing Germany, Fildes drew up contingency plans for dropping more than four million four-pound anthrax bombs on six major German cities: Berlin, Hamburg, Stuttgart, Frankfurt, Wilhelmshafen, and Aachen. The cities were to be attacked simultaneously by a heavy bomber force carrying forty thousand five-hundred-pound containers, each containing a cluster of 106 four-pound anthrax bombs. The intended result was a combination of inhalation anthrax from poisonous clouds and cutaneous anthrax from wound infections. Civilian German deaths were projected at three million.[31] Although the Americans were still having technical difficulties, Fildes saw no reason the difficulties and dangers could not be overcome. He projected the Vigo factory could produce 500,000 four-pound bombs per month and that in eight and a half months it could make the four million plus bombs estimated to obtain 50 percent fatality rates in the German urban targets, of which Berlin, at 228 square miles, was the largest. Fildes also reported that Porton had developed a second advantage, the agent for brucellosis, and that this agent could be even more easily produced and put in bombs, in the same US factory.

In the summer of 1944, Fildes and the Royal Air Force devised plans for aircraft returning from conventional bombing raids to drop the five million anthrax cattle cakes from Operation Vegetarian on Germany, with the aim of wiping out its beef and dairy herds. These plans were never carried out as the Germans were already being defeated by conventional arms.

Compared to those of the United Kingdom, US wartime resources were lavish and the organization and scale of mobilization was enormous. The Army and Navy were involved in SPD projects, along with the Departments of Agriculture and Animal Husbandry and the office of the Surgeon General, which limited its concerns to public health and safety issues. In February 1944, in preparation for the planned Allied offensive in Normandy,

Camp Detrick established a school for training biological warfare officers, who were then dispatched to all major military operations in Europe and the Pacific. Whereas the British gave instruction on biological weapons only to the highest command levels, the US military attempted to communicate with all officers above the troop level in the CWS and in American medical and intelligence services. This intelligence component, yielding feedback from far-flung sources, became one source for the discovery that the German biological weapons threat had been greatly exaggerated.[32]

Despite some personnel problems, the Americans integrated a large and educated workforce into the biological weapons program, far in excess of the British: "In its fourth year of work, the Biological Section at Porton under Dr. Paul Fildes was comprised of only 45 people, including 15 officer-civilians (four of whom were being supplied by SPD), 20 enlisted technicians, and 10 female helpers. In contrast, the year-old SPD had more than 1,500 people at Camp Detrick and Horn Island, the two installations operating in May 1944, and this figure was to be more than doubled in the next year."[33]

The SPD projects were also dispersed among many universities where more people were engaged. Harvard scientists developed vaccines against fowl diseases and a toxoid against botulinum toxin. The University of Chicago's Institute of Toxicology worked on aerosols, Northwestern University and the University of California at San Francisco on shellfish toxins, the University of Cincinnati and University of Kansas on tularemia, the University of Notre Dame on rickettsia, the Stanford University School of Medicine on the pathogenic fungus *Coccidioides*, Ohio State University and the University of Chicago on plant growth regulators (the precursors of Agent Orange), and the University of Wisconsin on the mass culture of spores. Cornell University had a project on anthrax, Michigan State University one on brucellosis.[34] The Navy, the Department of Agriculture, the National Institute of Health, and various public health departments all had laboratory projects connected with biological weapons.

The open spaces of the United States and Canada served as field test sites with a diversity the reflecting the ecologies of potential war zones—deserts, mountains, forests, the tropics—and their meteorological and weather patterns and possibilities for biological agents. At Horn Island, for example, the Navy conducted a series of experiments with flies and mosquitoes likely to be found in the Pacific Islands.

Back at Porton, Paul Fildes eventually became dissatisfied that the US government, instead of sticking with the UK agenda, had embarked on a broad spectrum of research, "treating [biological weapons] as a subject to be explored *de novo* and with special emphasis on the Pacific theatre."[35]

Military Control

Throughout 1943, the US military gained more control of biological weapons initiatives and had less use for civilian scientists. Ira Baldwin experienced this change firsthand at Vigo. By his account, the Army was in a great rush to get the Vigo plant into production, and its Corps of Engineers found Baldwin's conscientious onsite safety officer an obstacle to its agenda. Informed that the Army intended to replace the safety officer with a chemical engineer untrained in biological engineering, Baldwin resigned his position as technical director at Camp Detrick and became an advisor instead.[36]

Soon after, civilian science advisors, including Baldwin, became involved in an interpretation of intelligence data that nearly resulted in hundreds of thousands of Allied soldiers being unnecessarily inoculated against botulinum toxin. At a top-secret December 1943 conference in Washington, science advisors from the United Kingdom, Canada, and the United States were privy to a briefing that combined intelligence about a new German "flying bomb" (later known as the V-1 rocket) with technical advances SPD had made in the last year with bacteriological payloads.[37] To this was added the claim of a refugee German scientist, Helmuth Simons, that the Germans had developed botulinum toxin as a biological weapon. The OSS and other intelligence agencies had already dismissed Simons as an unreliable informant. The science advisors in attendance, focused on the lethality of botulinum toxin (code-named "X"), remained concerned the Germans might have produced it in bulk. Convinced of this threat, George Merck immediately asked Henry Stimson for permission to approve the manufacture and stockpiling of biological agents for retaliation. Stimson responded to Merck's alarm by pointing out that scientists should stick to experiments and let the military determine production needs. Stimson then shifted all biological weapons production-oriented research and development to the CWS, confining the WRS scientific advisors to peripheral plans and projects.

In January and February 1944, the advisors had another chance when a special War Department subcommittee (called the Barcelona Committee) convened to reconsider the German agent "X" threat. Its members included George Merck, Ira Baldwin, and Gen. Alden Waitt, now head of the CWS. The committee, emphasizing the botulinum-toxin rocket threat, called for immediately increasing US offensive and defensive capacity. The Canadian and US governments cooperated in a rush campaign to provide Allied troops with enough toxoid inoculations, in time for the Allied invasion planned for spring. The Canadians set about producing 150,000 doses of the experimental toxoid. Then, without informing the Canadians, the US government cancelled its toxoid preparations. The United States had received reliable information from Ultra (the code name for the Allied interception of German wireless messages) that the Germans had no biological weapons capacity and were much afraid of inadvertently provoking its use by the Allies. But the Canadian government at the highest level knew only that its troops could be in danger. Having prepared crates of toxoid for overseas shipment, it pressed for troop inoculations nearly until June 6, D-Day, when the plans were finally abandoned. The result was a loss of face for Canadian scientists E. G. D. Murray and Otto Maass, who were long left out of the inner circle of US decision making.

As the war continued and the SPD posts and projects grew, the military took more control and civilian advisors became increasingly marginal, which also happened in the United Kingdom. In August 1944, the War Service Division and its advisory subdivisions were dissolved and George Merck became a special consultant to Stimson. The rationale was that a civilian agency had been part of the secrecy the biological weapons program initially needed "for more perfect cover." That need was now superseded by the need for military development, planning, and preparation; therefore, "the responsibility for biological warfare should now be unified and centralized with the military establishment."[38]

Biological Warfare at the End of World War II

On August 11, 1945, three days before Japan surrendered, a memo went down from the War Department to the head of the CWS to end the SPD's work. Camp Detrick and Vigo were to be converted to peacetime purposes. At Camp Detrick the number of personnel quickly dropped from 2,273 to 865; most who left were enlisted men, and most who stayed were Army and Navy officers. The attrition was similar at Vigo. The Horn Island station was closed. So, too, was the one at Granite Peak, Utah, although there remained an agreement with Dugway officials for future biological weapons tests several months a year.

The decommissioning of the Vigo plant was a major enterprise. Nearly six hundred contracts were cancelled, including one for $4.3 million with Electromaster, Inc., in Detroit. Excess experimental animals were shipped to Edgewood Arsenal or the Department of Agriculture. Most of the military vehicles, bedding, and blankets went back to the Army. Machine shop equipment went to Camp Detrick, along with twenty thousand Mark I bombs. Other equipment was donated to universities, technical institutes, and to the federal penitentiary at Terre Haute. At the end of October, almost $800,000 worth of equipment had been declared excess. It took eighteen boxcars to remove the equipment from Vigo. The ten 20,000-gallon fermenters were left behind and later used by Charles A. Pfizer and Company, to whom the plant was leased and then sold.

The Vigo plant nearly achieved its goal and lent credence to any future industrial-scale project. In June and July 1945, trial runs with simulant *Bacillus globigii* demonstrated that the plant was "capable of producing 8000 pounds of a concentrated anthrax simulant slurry in a single run, under satisfactory conditions of procedure and with maximum safety to operating personnel."[39]

The 1945 attrition in personnel and retraction in projects made it seem the biological weapons program was gone for good. But enough scientists stayed to work for the Chemical Corps to revive the biological weapons program when, a few years after the war, its mission was redefined to meet Cold War threats.

US Accomplishments by War's End

By the end of the war, the US biological weapons program had cost $400 million, perhaps not much compared to the atom bomb project, which cost $2 billion, but this investment was historically unprecedented.[40] Most of the biological weapons expense was for research and development, not industrial production, as in most other wartime weapons programs.

In a September 1945 memo to the secretary of war, George Merck summarized the results: pilot and large-scale production plants for pathogens; mass production of anthrax and virulent brucellosis for bomb filling; the standardization and production of the four-pound Mark I bomb; improvement of its cluster adapter and container; the development and field testing of a new cluster bomb, the SS; the development of field test methods for biological munitions; pilot plant and large-scale production plants for anticrop agents, especially against rice; the discovery and use of crop-destroying and defoliating chemicals thought not to be injurious to humans or animals; and the development of toxoids to immunize against botulinum toxins A and B.[41]

The memo cited "great advances" in understanding the pathogenesis of airborne infectious agents and immunization against anthrax, tularemia, and brucellosis. Postexposure therapies (penicillin for anthrax, streptomycin for tularemia, and sulfadiazine for glanders) were other advances. The program produced a better understanding of pathogenic organisms, of airborne diseases, and of plant diseases. Vaccines were developed against two poultry diseases (Newcastle disease and fowl plague) and rinderpest, which later proved a boon in controlling the disease in China. For the first time, a pure bacterial toxin (botulinum toxin) had been isolated.

The US biological weapons facilities were the basis for developing technical defenses against dangerous pathogens, as well as safety standards. Laboratory workers needed protection, and for them Army engineers developed lightweight masks with near-zero leakage and special protective suits that are the prototypes for today's hazardous materials and the safety gear required in high containment laboratories. Air and environmental sampling kits were developed for testing in field or base laboratories near combat areas. Improved methods were such that the new infectious disease laboratory of the National Institute of Health at Bethesda, Maryland, was modeled after the laboratories and pilot plants at Camp Detrick.

In a classified Department of State report, Merck and three other officials of the biological weapons program were grim about future biological threats to national security. Enough had been learned to realize that biological weapons were inexpensive, that virulence could easily be enhanced, and that, compared to atomic weapons, it would be difficult to control research and development.[42] Others thought that although any resourceful nation could produce pathogens for sabotage, only nations with sizeable resources in scientific personnel and industrial facilities could develop and sustain biological weapons.[43]

In October 1944, two top science advisors, Vannevar Bush and James Conant, foresaw a biological weapons arms race unless the United States could dispel the secrecy surrounding the world monopoly it now possessed. In a letter to Stimson, they argued that the US government should take the lead in having all information about biological warfare placed under an international organization capable of control and inspections. In a second memo in February 1945, Bush argued for a United Nations effort to police aggression and to prevent any peace-loving nation from having "to fear the scientific activities of another."[44]

Stimson seems not to have communicated this idea to Roosevelt, who was seriously ill, or to have himself given it any serious consideration. Bush and Conant made the same argument to the Interim Committee on atomic energy. They had seen how the United States and United Kingdom excluded the Soviet Union from all information concerning the development and use of the atom bomb. Utopianist in their vision, they sought to integrate the USSR into a world community united against weapons of total war.

Neither appeal was successful. Within a few years, the Cold War revived the secret US biological weapons program, which looked to the annihilating capacity of nuclear weapons for definitions of its own destructive power.

CHAPTER 4

Secret Sharing and the Japanese Biological Weapons Program (1934–1945)

As World War II was ending, Secretary of War Henry Stimson ordered that research and development in biological warfare should continue. The Chief of the Chemical Warfare Service (renamed the Chemical Corps in 1946) was put in charge of the program, and he was to be guided by the advice of the Surgeon General on medical aspects. The Army and Navy were expected to continue their cooperation. The newly established Air Force joined them in 1948. Partnerships with Canada and the United Kingdom also continued.[1]

For a brief time after the war, the US program showed considerable openness about its wartime activities and their contributions to scientific research. In October 1945, after arguing that an open publication policy would retain high-quality biologists and attract new ones, the Chemical Warfare Service was allowed to relax its publication rules.[2] From then until 30 June 1947, 156 scientific papers passed approval at Camp Detrick. Of these, 121 were published in the open literature, 15 were accepted for journal publication, 20 articles were under review, and another 28 papers had been presented at professional meetings. The subjects covered included aerobiology, bacteriology, veterinary medicine, and the study of plant pathogens. In May 1947, the formerly secret report by Rosebury and Kabat discussed in chapter 1 was published by the *Journal of Immunology*. Rosebury also brought out his book *Experimental Air-Borne Infection* as a publication of the Society of American Bacteriologists. This work contained a diagram of an aerosol chamber devised at Camp Detrick, photographs of equipment, and seventy-five tables of information on test results involving the anthrax simulant *B. globigii*, the psittacosis virus, brucellosis, and tularemia.[3]

Simultaneously, in stark contrast to this openness, biological weapons scientists from the Chemical Corps were involved in a secret arrangement to procure technical information from Japanese biological warfare scientists in return for protecting them from prosecution for war crimes. In Manchuria, from 1934 through 1945, the Japanese had conducted human-subjects experiments, including vivisection, on thousands of Chinese prisoners and had attacked and killed civilians with biological agents. The immunity agreement, finalized in 1948, was a watershed event, a divide between retaliatory biological weapons as part of the World War II effort against the Axis powers and a new, Cold War vision of biological weapons as potentially comparable to the atomic bomb in destructive scope. The American bargain with the Japanese scientists helped reposition the US program as a top secret venture based on total war doctrine in the nuclear age.

Intelligence and the Immunity Bargain

In 1942, China presented the Allies with allegations that the Japanese Army had attacked its cities with plague-infected fleas, causing significant deadly outbreaks. The Western press ran the story, which raised suspicions, but then the allegations faded from the news.

With the defeat of Japan in 1945, the United States quickly sent scientists to investigate and appropriate innovative Japanese weapons technology, just as it had done earlier in Germany in what is known as the Alsos Mission.[4] The earliest scientific mission to Japan, in November 1945, included five civilian scientists and four members from the military, three from the Chemical Warfare Service.[5] After six weeks of interviews and site visits, investigators found little of interest in Japan's chemical program and even less in its attempt to develop atomic weapons. One of the American conclusions, though, communicated to President Harry Truman, was that the Japanese biological weapons program might add to American military knowledge. Lt. Col. Murray Sanders, one of the CWS members of the group, quickly wrote a report summarizing Japanese accomplishments in bacteriology; this report was followed by another from the Army in 1946, written by Lt. Col. Arvo Thompson, which credited the Japanese program with developing effective vaccines and curative sera for plague, typhoid, dysentery, and other diseases, as well as diagnostic serums and antigens and drugs, including penicillin.[6]

Japanese biological warfare scientists also reported having done research on animal diseases, including anthrax and glanders, and on anticrop agents. When questioned, the Japanese scientists denied to the American mission that they had prepared any biological weapons for offensive purposes or that they had engaged in experimentation on humans.

In early 1947, US military intelligence (G-2) agents in Tokyo invited Dr. Norbert Fell, a division chief from Camp Detrick, to reevaluate the Japanese program. He was not to estimate whether criminal prosecution was warranted, but to find out if its scientists had any valuable information that the United States should acquire for its national security. Gen. Ishii Shiro (1892–1959), the military biologist who had created the program, had been located at his home in the Tokyo suburbs; with him came the promise of important hidden documents being revealed. Might this information be worth some kind of negotiation? The same postwar question had been posed regarding Nazi Germany's weapons scientists, with whom deals had been struck for amnesty, immigration, and US defense employment.[7] Those arrangements, as far as is known, did not involve war criminals who had engaged in brutal human experiments on captives, which the Japanese military scientists, despite their denials, had conducted.

G-2 officials appeared to leave the evaluation of scientific data, key to an immunity bargain, to Norbert Fell and several other CWS experts who also came to Japan. But by the time Fell arrived in Tokyo, in May 1947, US intelligence had already laid significant groundwork for a secret exchange. In late 1944, Secretary Stimson was told of the likelihood of a Japanese biological warfare program. In the following year, military intelligence gathered more supporting data, although the threat of biological weapons to the United States in the Pacific Theater seemed minimal. At the time, Washington was also interested in keeping the subject of biological weapons out of the news, lest the secret US program become public knowledge.[8] In September 1945, the Office of Strategic Services (OSS), the forerunner of the CIA, concluded that the Japanese had "perhaps the best informed scientists in BW investigations of any nation in the world" and that they had posed a significant threat to the United States.[9] Concerning the Japanese scientific activities, a G-2 report from January 1947 commented: "Naturally, the results of these experiments are of the highest intelligence value."[10] G-2 also managed to create a roadblock against the US Adjutant General's Office in Tokyo, which was tracking war criminals for prosecution. In March 1947, the Joint Chiefs of Staff sent an order placing G-2 in total charge of the biological warfare inquiry and mandating as essential "the utmost secrecy...in order to protect the interests of the United States and to guard against embarrassment."[11]

To US leaders, the reconstruction of Japan as a stable democracy was crucial to offsetting the Soviet presence in Asia and the rise of Communist China. For Japan to become a credible ally, extreme Japanese aggression during the war, including the inhumane treatment of prisoners of war and civilians in the Pacific, had to be quickly adjudicated and forgotten. Many top Japanese military and civil leaders were already under indictment for war crimes. If rumors about the Japanese biological warfare program were true, the addition of prosecution for human experiments akin to those practiced in Nazi death camps would have added another obstacle to the goal of reconstruction.

In addition, Ishii's program might have had the patronage of the Japanese emperor, himself a biologist and known to take an interest in the details of government administration. The American government, and particularly Gen. Douglas MacArthur, who was in charge of the occupation and reconstruction of Japan, had decided that the divine persona of the emperor had to be conserved or the entire hierarchical structure of Japanese society could fall into chaos and perhaps cause violence against occupying troops.[12] MacArthur, who controlled the war crimes trials, exempted the emperor from being prosecuted. Nevertheless, what the Japanese scientists might have revealed about his complicity in the biological warfare program could have been ruinous.

Entire networks of influence might have been implicated had the Japanese scientists been put on trial. Ishii had received support from the highest Japanese military and government officials. At least six of this cohort had already been indicted for prosecution in the Tokyo war crimes trials. Prime Minister Tojo Hideki, for example, when a high officer in the Kwantung Army (Japan's military force in Manchuria), had reportedly seen films of Ishii's human experiments; during the 1930s he had personally approved the expansion of the program.[13] Gen. Umezu Yoshijiro, chief of the Army General Staff and a signatory to the 1945 instrument of surrender that ended the Pacific War, had been a Kwantung commanding officer familiar with Ishii's program.[14] Officials not indicted had probably also supported the program.

Ishii was well connected in the medical world as well, which continually lent support to his program. Over the years, using those contacts, he had recruited hundreds of young medical students and technicians who circulated through its research facilities and were more or less complicit in the experiments. Sworn to secrecy at that time, after the war these educated people were poised to take positions in the new Japan's universities and industries.[15] If the program were judged a war crime, blame might go beyond the military indictments to professional civilian networks, which would suffer guilt by association.

In May and June 1947, Norbert Fell conducted interviews with Ishii, the visionary of the Japanese biological weapons program. With a small cadre of his colleagues, Fell concluded that what the Japanese had to offer was worth the trade for immunity. Mainly the material was firsthand accounts of a large, long-term weapons program and documents of medical experiments on humans (including some eight thousand microscopic slides and detailed autopsy data). The autopsies included those of victims of inhalational anthrax, of special interest to the Americans. A supplement to Fell's report noted that "such information could not be obtained in our own laboratories because of scruples attached to human experimentation."[16]

In Washington, the US government agencies that were consulted about the immunity bargain had slightly differing perspectives. State Department officials pointed to the risk of US government embarrassment should the agreement become public knowledge. In September 1946 the International Military Tribunal in Nuremberg had defined similar experiments by Nazi scientists as war crimes. G-2 and the Chemical Corps officials argued that the Japanese technical information was vital for national security and could not be made public at a trial or the Soviets would acquire it. Indeed, the Soviets had asked for access to Ishii and the other scientists, which the US government granted in 1947, after coaching the Japanese to keep their responses minimal. The Washington consensus was that national security took priority over criminal prosecution. Apparently Gen. MacArthur personally intervened to keep the Japanese biological warfare scientists from exposure.[17]

Exactly who authorized the immunity bargain remains unclear. Official approval should have come from the highest level of US government, from President Truman, Secretary of Defense James Forrestal, or Secretary of State George Marshall. A memo of 24 June 1947 by Fell identifies Truman and MacArthur, along with Alden Waitt, the Chemical Corps commander, as recommending the decision that "all information obtained in this investigation would be held in intelligence channels and not used for 'War Crimes' programs."[18] By the time the agreement was signed, in March 1948, the Tokyo war crimes trials had concluded without revelations about the atrocities committed by Japan's scientists involved in the biological program. The cooperating Japanese scientists and many others who had participated in the program were now free. Except for Ishii, who was not in good health, the dozen or so who directly benefited from the agreement became medical school professors and administrators or pursued successful careers in research or industry.[19]

Not all the Japanese military involved in biological weapons were as fortunate as Ishii and his associates. In 1949, a dozen captured by the Soviet Union as they attempted to flee China were put on trial in Kharbarovsk, one of four Siberian cities that Ishii had planned to attack late in the war. The five days of testimony reportedly attracted large crowds and had to be broadcast by loudspeakers outside the courthouse. The summary version of the Khabarovsk trial appeared in English in 1950.[20] Although the Khabarovsk testimony described human experiments already known to American and British officials, the US and UK governments dismissed the proceedings as Stalinist propaganda. The defendants were imprisoned, and by 1956 most were repatriated to Japan.

To Western intelligence, the short prison sentences given the Japanese defendants indicated that the Soviets had contracted a bargain for information not entirely unlike the American one and for similar purposes. US military intelligence had acquired what seemed reliable information on past Soviet biological weapons activities from a German chemical weapons scientist.[21] It suspected that the USSR had researched biological weapons during the 1930s and into the war years and could soon have the scientific and industrial resources to create a strategic threat.

The Modernist Context: Manchuria

The social and political context for the Japanese biological warfare program was imperial expansion into China. On 19 September 1931, in what became known as the "Manchurian incident," the Kwantung Army seized the pretext of being attacked to begin its takeover of all key Manchurian cities along vital rail lines. Within six months, with help from jingoistic media and ultranationalist leaders, the Japanese public succumbed to "war fever" and began to define Manchuria as a "lifeline" essential to its national survival.[22] In early 1932, the Japanese recreated Manchuria as Manchukuo, a puppet state, with Henry Pu Yi, the last Chinese emperor, as ruler. In 1933, Japan resigned from the League of Nations following worldwide protest.

Manchukuo was planned as a modern utopia with new cities based on science, culture, and industry. Japanese intellectuals and businessmen, seeking freedom from the old imperialism, took advantage of this social experiment, which they wanted to be "as modern as Soviet planning, Italian corporatism, German State socialism, and the American New Deal."[23] Among other visionaries, the inventor of the Japanese corporatist state, Kishi Nobusuke, used Manchukuo as a test ground for fascist economic policies.[24]

As its expansionist objectives were being fulfilled in Manchukuo, Japan faced a military threat from the Soviet Union, its major military adversary. Despite international agreements and a neutrality pact designed to minimize friction, the Japanese chronically chafed at sharing Asian borders with the USSR. In 1939 open warfare broke out between the two countries. In 1941, Japan and the Soviet Union concluded another neutrality pact, although the Japanese were still wary of the Soviets.

Outside Manchukuo, Chinese Nationalist forces were in conflict with Chinese Communists, and both governments in the civil war were hostile to Japan. In 1937 a border skirmish not unlike the Manchurian incident gave Japan the opportunity to enter the China war.[25] Japan's military superiority was unquestionable. So too was its willingness to brutalize Chinese civilians, as in the 1937 Rape of Nanking. Japanese dominion soon extended along the eastern coast of China, south to Canton and beyond.

Before this war, while Manchukuo developed as a Japanese "paradise" with new cities and industries, tight, often ferocious military and police control was exerted over Chinese insurgents, guerrillas, and so-called bandits who were the source of "experimental material" for Ishii's program.[26] The Kwantung Army had four non-Japanese populations to subordinate: the majority Han Chinese who had immigrated from the south, the indigenous Manchurians, Koreans imported as labor, and Russians and other European émigrés. All of these groups were considered racially and culturally inferior, especially the Han Chinese. The army was rarely held responsible by Tokyo for its methods; using its military police, it kept order by its own frontier rules. Manchukuo was thus an optimal environment for a secret military program that could evade legal and moral accountability, attract scientists with the promise of good pay and research opportunities, and conduct human experiments on a despised civilian population.

The Military Doctor as Visionary

As early as 1927, Ishii was making the case to his superiors for a biological weapons program.[27] Part of his reasoning was that, after signing the 1925 Geneva Protocol, most nations in the world would forgo biological weapons, giving Japan (a signatory but not then a party to the treaty) the outsider's advantage in developing them. Western biological science, already promoted in Japanese universities and medical schools, promised to bring modern efficiency to the Japanese arsenal. Biological weapons required no very large industrial base and they were relatively easy to produce. In 1928, Ishii went on a two-year world tour to pursue his interest in bacteriology. As a junior officer, he had achieved some fame for inventing a portable water-filtration system for mobile troops, even demonstrating this device to the emperor. His goal on his trip abroad was to learn more about the latest in Western biology; using diplomatic letters of introduction, he visited research laboratories in Germany, France, the Soviet Union, and America.[28]

On returning to Japan, Ishii was appointed professor of immunology at the prestigious Tokyo Army Medical School and promoted to major. In his absence, the 1929 economic crash had brought an end to Japanese liberalism and a return to isolationist politics. The military, in wresting control of the government, was turning to Nazi Germany and Fascist Italy as exemplars and allies. As Ishii lobbied for a biological program, he enjoyed personal support from the highest military levels, especially among ultranationalists, and from the head of the Ministry of Public Health.[29] Soon Ishii was given his own laboratory building at the medical school. There he drew a distinction between highly secret "A" research on offensive weapons and "B" research on defensive ones.

In 1932, his drive to do "A" research in Tokyo presented security problems, and Ishii was given permission to relocate in Manchukuo, where he had the full cooperation of the Kwantung Army. The difficulty of maintaining secrecy at his first installation, in the northern city of Harbin, led to the construction of a large walled compound one hundred kilometers to the south, the Zhong Ma Prison Camp. The camp was built to house between five hundred and one thousand inmates. Two years later, when a cohort of escaped prisoners blew up the facility, Ishii's military superiors in Tokyo immediately agreed to fund a new and larger research compound. Generous support for the biological weapons program continued until the end of the war in 1945.

Ishii's new, heavily guarded enterprise was built on the site of a cluster of ten villages, known collectively as Ping Fan, twenty-four kilometers south of Harbin. When completed in 1939, it consisted of seventy concrete buildings, including laboratory and pharmaceutical facilities, three crematoria, several unheated outbuildings, and an airfield for Ishii's private fleet of biplanes. Intense secrecy surrounded the program. For the Japanese who worked there, mail was censored and travel was restricted. Unmarked trucks delivered material and equipment. In 1941, Ping Fan was renamed Unit 731. Anda, north of Harbin, was the Unit 731 outdoor field area for bomb tests, which also used human subjects.

In 1936 the Japanese signed the Anti-Comintern Pact with the Germans and Italians, thereby finally rejecting its old alliance with the British and their presence in Asia, where Japan intended to rule supreme. The year 1936 also marked the construction of Unit 100, named the Kwantung Army Anti-Epizootic Protection of Horses Unit, headed by Wakamatsu Yujiro, a military veterinarian who was also granted immunity by the US government in 1948. Located six kilometers south of Changchun, the new capital of Manchukuo, Wakamatsu's ultrasecret facility was nearly the size of Ping Fan.

Unit 731 and Unit 100 were at the center of a system of research satellites that eventually extended to other Manchukuo and Chinese cities. Japanese academic institutions and hospitals were also involved, including the universities at Kyoto (where Ishii had gone to medical school) and Tokyo. From these and other institutions Ishii regularly recruited young scientists and technicians with promises of high wages and innovative experiments. In 1939, Ei Unit 1644, a satellite of Unit 731, was added to the program in Nanking. During World War II, as Japanese control expanded from the eastern coast of China into French Indochina, Siam, and Burma, other biological weapons satellites were established as outposts.

The extent to which any of these smaller centers engaged in human experiments, as opposed to basic laboratory research or vaccine production, remains unknown. Defensive work was also conducted at Unit 731. It manufactured twenty million doses of vaccine a year, most of it a typhus vaccine. Research on vaccines for anthrax, plague, gas gangrene, tetanus, cholera, and dysentery were also part of the program's agenda, along with research on blood products to treat a variety of diseases and for diagnosis. As for volume pathogen production, the Japanese lagged behind the industrial levels achieved, for example, by the Camp Detrick pilot plants for anthrax.

Unit 731, Unit 100, and the other sites in Manchuria and China were blown up by the Japanese before they fled in 1945. Documents were also destroyed, which made the information saved by Ishii and others that more valuable.

Japanese Biological Weapons Activities

Camp Detrick representatives were keenly interested in any mechanical delivery systems (bombs, artillery shells, or spray devices) that Ishii might have developed and in reports of actual biological attacks. The Americans ultimately found they had little to learn. The Japanese attempted but were unable to invent any bombs that were practical; although nine different types were developed, along with a prototype bomb for anthrax spores, called the HA bomb, none had effective dispersal capacity.[30] Under Ishii's direction, over a period of five or six years, more than two thousand steel-walled bombs were detonated in field tests using humans who were bound in place at the site.

Many of the victims were killed by shrapnel wounds rather than biological pathogens. Ishii also experimented unsuccessfully with spray bacteria from low-flying planes. He had the intent but not the technology to link biological weapons to the airplane and strategic attack plans. Some joint experiments were held with the Japanese chemical program, apparently also involving human victims.[31] But in general the chemical program was a separate venture, unlike in the West, where chemical weapons expertise and technology aided innovations in biological weapons.

Thus limited, Ishii devised technologically unsophisticated tactics that were more like disruptive sabotage. One was to lure enemy troops into a former battlefield secretly contaminated with infectious disease. Another was to inundate a civilian area with plague-infected fleas, either dropped by small low-flying planes or scattered by spies.

In 1939, Ishii was given the chance to use his bacteriological weapons. Throughout that summer, the Japanese attempted to gain control from the USSR of the Nomonhan border, a complex area near the frontiers of Mongolia, Inner Mongolia, Manchuria, and the Soviet Union. In August (a week after the USSR and Germany signed their Non-Aggression Pact), Soviet Marshall Georgiy Zhukov led three divisions and five mechanized brigades to the border. There they encircled and routed the Japanese, who lost eleven thousand of their fifteen thousand soldiers in the conflict. As the Japanese retreated, a suicide detachment trained by Ishii stayed behind to infect the Khalkhin-Gol River with typhus, paratyphus, cholera, and dysentery. Whether these attempts had any effect on the Soviet soldiers was undetermined.

In 1940, Ishii organized five months of sporadic attacks on the southern port city of Ning Bo in Chekiang (Zhejiang) Province, in an attempt to disrupt food transports for Chiang Kai Shek (Jiang Jieshi), the leader of Nationalist China. Ishii used fleas to start plague outbreaks and attempted to infect water sources with typhoid and cholera germs. Japanese airplanes disseminated contaminated wheat and millet. Reportedly, a thousand people were made sick, with one hundred fatalities in the city and region. Ning Bo and nearby towns were affected by plague outbreaks the following year and again in 1946 and 1947. Some shipping and food transport may have been disrupted by these attempts, but under the general conditions of war, the Japanese found this effect difficult to measure.

In 1942 Ishii became involved in a difficult military campaign waged broadly through Chekiang Province. For example, in and around the cities of Yushan, Kinhwa, and Futsing, while the Japanese army staged a strategic retreat, three hundred of Ishii's men stayed behind to spray germs and infect water supplies, and three planes under fighter escort dropped bacteria and plague-infected fleas. Then, it was hoped, unsuspecting Chinese forces would invade the area and be overwhelmed by epidemics of cholera, dysentery, anthrax, plague, typhoid, and paratyphoid. Although Ishii later claimed that this attack was a great success, it was more likely a disaster. Ignorant of the germ attack, many Japanese soldiers apparently were exposed and as many as a thousand died. Shortly after, Ishii's authority was curtailed; he was removed as head of Unit 731 and transferred to the Nanking post.

The Japanese army also experimented with food and water contamination. In the Chekiang campaign, three thousand Chinese prisoners of war were given food contaminated with typhoid and paratyphoid bacteria and then released to spread the disease unknowingly as they returned home. This food distribution (but not its aftermath) was filmed and shown in Japan as an example of Kwantung Army beneficence to Chinese peasants. Japanese soldiers were also given contaminated food to leave behind along roadsides for hungry Chinese civilians or soldiers to find.

In exchange for immunity, in addition to their laboratory records, Japanese biological warfare scientists reported to US military scientists and intelligence officials the details of twelve field "trials" in which they used biological weapons, including those just described, plus other plague attacks on Chinese cities and towns.[32] One advantage for the Japanese, which Ishii understood, was that in occupied territory and wartime China their germ weapon attacks could also be mistaken for natural outbreaks. This was one rationale for using plague-infected fleas.

As the war was ending and Japan's position grew desperate, Ishii attempted to convince top Japanese military officers to resort to biological warfare. His superiors generally resisted his plans.[33] Yet earlier, in 1941, the Japanese army reportedly considered biological attacks on the Philippines in 1941, and in 1942 it listed Australia, India, and Samoa as potential targets, and then Burma and New Guinea.[34] The concept of total war was not foreign to Ishii. His chief mentor, the powerful Gen. Nagata Tetsuzan, was a strong advocate of military modernization with strategic objectives that included civilian targets.[35]

In 1947, after reviewing some sixty pages of Japanese material, the report by Detrick microbiologist Norbert Fell concluded that in the conduct of biological warfare the Japanese had little to offer, but he was optimistic about the human experimentation data: "It is evident that we were well ahead of the Japanese in production on a large scale, in meteorological research, and in practical munitions... However, data on human experiments, when we have correlated it with data we and our Allies have on animals, may prove invaluable, and the pathological studies and other information about human diseases may help materially in our attempts at developing really effective vaccines for anthrax, plague, and glanders."[36]

Biological Weapons Agents: What Was Learned?

Fell's optimism about the promised Japanese human experiment data must have been short-lived. Ishii, who controlled Unit 731 and Unit Ei 622, took a scattershot approach to investigating a fairly wide range of biological agents. Physician Edwin Hill, the Chief of Basic Sciences at Detrick sent to Tokyo in 1947, reported back to CWS commander Waitt a list of twenty diseases explored within the program, plus unspecified anticrop agents.[37]

Ultimately, the Japanese scientists presented the Americans with autopsy material in twenty different disease categories with a total of 859 human cases. Of these, 394 were categorized by them as "adequate" for analysis. The majority of deaths were attributed to cholera (50), Korean or Songo hemorrhagic fever (52), tuberculosis (41), and bubonic plague (106). (The Japanese divided the plague autopsy cases into 64 "epidemic" and 42 laboratory cases.) As Fell wrote in his assessment: "In general it was concluded that the only two effective B.W. agents they had studied were anthrax (and this agent was considered mainly useful against livestock) and the plague-infected flea. The Japanese were not even satisfied with these agents because they thought it would be fairly easy to immunize against them."[38]

It was under Dr. Wakamatsu at Unit 100 that these two diseases, anthrax and plague, were most thoroughly researched. Using animal research, US and UK scientists had developed an anthrax aerosol and considered it the most potent means of killing an enemy. Since the human inhalational form of the disease was rare, though, they kept returning to the problem of dose response and the difficulty of knowing how the disease progressed. On appearance, Wakamatsu and his colleagues were offering the Americans a rare prize: twenty-two autopsy records of human inhalational anthrax cases, among men from age twenty-five to thirty-seven, complete with color drawings of the fatal progression of the disease.[39]

The Camp Detrick scientists noted that such a small number of cases had no statistical value. Another obvious drawback was that the experimental conditions under which the anthrax victims were fatally infected represented nothing that would occur in battle or in mass attacks on civilians. Lacking a cloud chamber or techniques to create a fine anthrax aerosol, the Japanese resorted to putting four men at a time in a small glass chamber fitted with spray pumps operated from outside. High volumes of crude anthrax spore suspensions were then pumped directly at the victims' faces. The victims first developed tonsillitis and bronchial lesions, perhaps from sheer inundation with anthrax spores. After that, the infection spread quickly. The outcome for all was rapid death, two to four days after exposure.

Biological Warfare Scientists

The research activities of the Japanese biological warfare scientists had much in common with those of their contemporaries, Nazi Germany's biologists and physicians, who also conducted inhumane experiments on captives.[40] The motivations for American biological warfare scientists to aid in protecting Japanese war criminals can only be guessed. They obviously placed national security above ethics and justice. Much later Murray Sanders said the immunity bargain was meant to protect the emperor's authority for the immediate purpose of preventing Japanese violence against occupying American soldiers.[41]

The future of the US biological weapons program might also have been at stake. The postwar program could have been permanently reduced to defensive work with vaccines and masks, unless technological progress for offensive purposes could be imagined, against a substantial political enemy. Working in concert with G-2, the Detrick scientists were apprised of the Soviet threat and positioned themselves to respond. On 24 June 1947, Norbert Fell reported to military intelligence through his superior, Commander Waitt: "The information that has been received so far is proving of great interest here and will have a great deal of value in the future development of our program."[42] Fell then indicated that important data was forthcoming from the Japanese, for example, a report on mathematical calculations of aerosol dispersal.

Perhaps, too, the Camp Detrick scientists were covetous of a kind of forbidden fruit, the human research information that their scruples prevented them from acquiring.[43] They could not have conceivably conducted the kinds of experiments on anthrax, tularemia, glanders, and other diseases that the Japanese had spent years performing. Nor had the Americans crossed the line to using biological weapons the way the Japanese had done, however primitive their methods.

Due to the reliance of Japanese medicine on Western and especially German science, the Japanese and the American scientists had a professional culture in common, which might have promoted some empathy. Despite the language barrier, the Japanese use of Western laboratory and autopsy techniques, materials, and descriptive scientific terminology proved easily comprehensible to their American counterparts. Japanese autopsy reports, for example, used German terminology characteristic of textbooks assigned in US medical schools. One of the shared scientific norms was a clinical detachment from the suffering and death of the human subjects. In the available US documents, American biological weapons experts appear to accept the Japanese human experimentation data without moral revulsion, as if it were clinical science.

More narrowly, Dr. Fell and other emissaries from Camp Detrick shared a particular wartime military culture with their Japanese counterparts, whose secret program aims had matched theirs. Perfecting biological bombs, determining infectious dose levels, developing battlefield and sabotage use of disease, planning mass attacks on civilians—both sides had pursued these goals in the interests of national defense. Professional deference was accorded the Japanese scientists, with whom the Americans from Detrick and G-2 had tea and dined. Japan was no longer the enemy; the enemy was the Soviet Union. On 12 December 1947, Edwin Hill enthusiastically noted that the Japanese reports were a bargain, costing relatively little to retrieve and process, given the years the well-funded program endured. He added, "It is hoped that the individuals who voluntarily contributed this information will be spared embarrassment because of it and that every effort will be made to prevent this information from falling into other hands."[44] To his report, Fell appended the collegial sentiment that he and the other American biological warfare scientists were looking forward to Ishii's memoirs on his military career. Yet he and the other American scientists must have realized that those memoirs, if honest, would have been confessions of criminality.

In the process of negotiation and information disclosure, Ishii and his cohort quickly realized that the Americans were no more interested in exposing the Japanese program than they were. The American government wanted to protect the emperor and keep information from the Soviet Union. It was perhaps equally important for the continuation of the Camp Detrick program to keep the American public in the dark. The years 1945–47 marked a brief period of openness about US defensive wartime research on biological weapons, and that information release alone created more media attention than the Army wanted. The grotesque details of the Japanese program could well have tainted the Camp Detrick program and restrained the governments of the United States and the United Kingdom (which was privy to the Japanese bargain) from continuing their offensive projects. After protecting SPD during the war, US intelligence continued to protect the biological warfare program after the war and into the Cold War, when the CIA developed a client's interest in biological weapons agents. In 1948, when the Japanese immunity bargain was completed, the head of the CWS and the US biological warfare scientists became complicit with US intelligence in sustaining secrecy around the criminal violations in the Japanese program. Together, US government officials and the military scientists involved in the Japanese immunity bargain placed national security, as it was interpreted at the time, above the law and above morality. Soon after the bargain was struck, as the Cold War intensified, secrecy descended on the US biological weapons program.

Secrecy Breakdown

After the war, it was known that the Japanese Imperial Army had established a biological weapons program. Yet details about the program—the human experimentation information and the treatment of prisoners—were not well known to the public, and for years the governments of Japan, the United States, and the United Kingdom denied them. The Japanese attacks on Chinese cities and towns were given short shrift, as if the Chinese were untrustworthy reporters of the events. Corroboration from Japanese scientists who had the evidence to prove the Chinese right would have corrected that perception, but their testimony and reports were kept secret.

Two decades after the Japanese biological warfare program terminated, attention to its criminal aspects began emerging sporadically in the media. In 1976 the Tokyo Broadcasting System televised a documentary, *A Bruise—Terror of the 731 Corps*, shown in Europe but not the United States. Five former biological warfare scientists revealed how they received immunity and how all their documents had been given to the United States. The story was carried by *The Washington Post* but caused no great sensation. [45] In 1981, an article on the program in the *Bulletin of the Atomic Scientists* by publisher John Powell attracted wider media attention. Powell had obtained documents on the Japanese program from US government archives, and he also relied on the Khabarovsk trial transcripts.[46] That same year, Tsuneishi Kei'ichi of Nagasaki University published *Kieta Saikinsen Butai* (in English, *The Germ Warfare Unit That Disappeared*),[47] based on his archival and interview research. Tsuneishi later assisted two British journalists, Peter Williams and David Wallace, in research for a 1985 Independent Television feature, *Unit 731—Did the Emperor Know?*[48]

Over the years, various Chinese and Japanese witnesses to the activities of the Japanese program made public statements. Some Japanese former scientists, in addition to confirming the worst of the human experiments, tried to explain the racist nationalism behind them. Said one, "I had already gotten to where I lacked pity. After all, we were implanted with a narrow racism, in the form of a belief in the superiority of the so-called Yamato [national Japanese] Race. We disparaged all other races... If we didn't have a feeling of racial superiority, we couldn't have done it."[49]

Sheldon Harris, whose 1994 book *Factories of Death* deepened the investigation of Unit 731, pointed to a persistent US and Japanese reluctance to release documents relating to the program.[50] In 2000 the US Congress passed the Japanese Imperial Army Disclosure Act, requiring "full disclosure of classified records and documents in the possession of the US government regarding the activity of the Japanese Imperial Army during World War II."[51] The Japanese government, with only weak laws supporting public access to documents, remained silent. In 2000 the Japanese court that heard the case brought by the families of Chinese victims admitted that great wrongs had been committed, but asserted that compensation was impossible because the 1951 San Francisco Treaty had put an end to compensation claims against Japan. Meanwhile, international scholarly investigation of the Japanese program continued and new material has emerged, at times unexpectedly. For example, also in 2000, around 450 pounds of program documents left by retreating Japanese were unearthed from a bunker in Hailar, near the Soviet border.[52]

Secrecy and the Moral Deficit

The immunity bargain had profound implications. For one, it prevented open international consideration of whether biological weapons programs, patently anticivilian, should be altogether banned. Policies regarding chemical and nuclear weapons were repeatedly exposed to public review and debate, in great measure because the consequences of their use were historically known. The effects of chemicals in trench warfare in World War I were well documented. After the bombings of Hiroshima and Nagasaki, judgments were possible about atomic weapons—their strategic worth, their deterrence value, and their danger to civilians. The Japanese state program was the first evidence of modern biological weapons use, however crude, and of the scientific methods that, although taken to extremes, were consistent with the objective of strategic attack. In the absence of historical examples, public debate about these weapons failed to take place. In 1947, some members of the American public might have favored keeping a biological weapons program and others protested. We will never know. The decision rested solely with a few military authorities and those in government who protected the program's secrecy and guaranteed its future expansion.

A legal and ethical deficit was also incurred, insofar as knowledge about Ishii's murderous program could have strengthened norms against state aggression on civilians. The Nuremberg trials and other criminal proceedings and the testimony of thousands of victims laid bare the structure and ideology of the Nazi state, with its industrial-age genocide in the name of racial purity. Once the world knew the consequences of Nazi Germany's racialist extermination policies, a moral lesson was learned and laws against genocide could be developed. Similarly, the realities of the devastation to Japanese civilians caused by US atomic bombs in 1945 contributed to restraints on nuclear weapons proliferation. In contrast, the Tokyo war crimes trials did not reveal how a civilized, technologically advanced society could for years conduct inhumane biological weapons experiments and attacks. The 1948 immunity bargain forfeited the important chance for more open discussion of the menace of biological weapons to civilians and the need to strengthen international law and other constraints against them.

The postwar history of legal restraints on biological weapons became profoundly troubled by government secrecy and a loss of public accountability, features that defined the United States program for the next two decades. By keeping the dreadful secrets of the Japanese biological warfare program, the US government lent legitimacy to offensive biological weapons aimed against civilian populations. John Powell described this consequence of the immunity bargain: "An 'insidious weapon' in enemy hands, biological warfare (BW) was transformed into an acceptable and valuable military tool when added to the American arsenal."[53] The military value of biological weapons had still to be judged, but there was no question that its program would continue to have a secure niche in state government.

While Fell and others were interviewing Japanese scientists, President Truman signed the 1947 National Security Act, a reorganization of the military services and intelligence organizations in reaction to the bombing of Pearl Harbor and the perceived threat of a Soviet-backed communist takeover of the world. Central to this reform was the threatening prospect of the Soviet Union as a nuclear power. Nuclear weapons would set the standard for the next twenty years of biological weapons development, making it imperative for biological warfare scientists to show how pathogens could devastate populations at the same enormous scale.

CHAPTER 5

Aiming for Nuclear Scale

The Cold War and the US Biological Warfare Program

The American officers who interviewed Japanese biological warfare scientists in 1947 quickly realized that US achievements were technically superior to those of Gen. Ishii Shiro's program. By the end of World War II, the American, British, and Canadian programs had learned much about airborne infection and about a diversity of biological agents, had improved defenses against some of them, and had come close to mass production of an anthrax bomb.[1] The atomic bomb and the Cold War signaled a momentous change in the US biological weapons program. The vision of the scale of intentionally spreading disease expanded to strategic attacks on a par with the destruction of Hiroshima and Nagasaki and with the Soviet Union and its allies as potential targets. For more than twenty years after the end of World War II, advocates of germ weapons struggled toward the goal of strategic potential and sought to convince the armed forces, especially the Air Force and Navy, that these new weapons should be assimilated. Few in the US government knew about this struggle for expanded offensive capacity. Within the program, research intensified, production capacity increased, and simulations of disease attacks on civilians grew larger and more elaborate until they verged on reality.

During the time the US postwar biological weapons program was being secretly ratcheted up to approximate nuclear scale, members of the American and international public, including scientists, were openly debating the dangers of nuclear weapons. Eventually, especially during the 1960s, the US chemical weapons program and the dangers it posed to civilians and the environment became a source of great contention. Yet, after just a brief flurry of publicity about US wartime research, biological weapons disappeared from public discussion for two decades. In 1947, Dwight Eisenhower, then Army Chief of Staff, banned disclosures regarding the program's continued activities. Following this, in a 1949 press statement, Secretary of Defense James Forrestal emphasized only US defensive research on biological weapons. His point, which the public seemed to accept, was that the biological program existed to defend the nation against the Soviet threat.[2] What the public did not realize was that this program for defense was squarely based on offensive goals. Since the United States was not bound by the Geneva Protocol, military advocates for biological and chemical weapons set about campaigning for the right of first use, which would allow the development of a fully offensive biological weapons program.

A 1951 report from the office of the Joint Chiefs of Staff (JCS) summarized the perceived military potential of biological warfare, including strategic and sabotage uses.[3] First, the report noted that the relative cheapness of biological weapons, compared to conventional, chemical, and nuclear arms, was a major benefit. The concern that lesser nations might acquire these weapons, which became central later in the century, was lacking; instead the presumption was that only industrial powers had the resources for biological weapons, as they did for nuclear and, in general, for chemicals. Thus, the JCS report concentrated on the USSR as an adversary and presented a number of World War II–type scenarios drawn from both the European and Asian theaters. In one scenario, the so-called nonlethal diseases (such as brucellosis) were suggested as a follow-up to atomic bombing in order to further debilitate enemy civilians in the area of attack. The use of biological agents in special operations was especially recommended, with civilians as unsuspecting targets: "Probably one of the most attractive and profitable uses to which BW could be placed would be in the field of covert operations. The ability of special agents or guerrilla forces operating behind enemy lines to place small quantities of BW agents accurately where they could be most effective offers tremendous possibilities. Covert use has a further advantage in that distinction between this type of BW and naturally occurring outbreaks is most difficult, and therefore might be used with crippling results well in advance of the initiation of hostilities."[4]

On a technical level, the first demand of the military was that this new kind of weapon must be dependably destructive. Commanders had only one goal, to win wars, and they were reluctant to experiment with germ weapons in battle. A 1952 JCS memo expressed this essential requirement and went further to indicate two important directives. One was that the Army's Chemical Corps should conduct field tests of biological weapons and the other was that each of the three services should prepare to accept biological weapons through the education or "indoctrination" of personnel:

> The Joint Chiefs of Staff desire to stress the fact that, regardless of the potential value in various situations of the available BW weapons, these will not be used unless the operational commander has confidence in the superior ability of the weapon concerned to contribute to the success of his mission. Therefore, tests to determine the operational capabilities of BW systems, and indoctrination measures to disseminate and create confidence in the test results, are necessary. Subject to coordination by the Secretary of the Army, who has been assigned responsibility for coordination of the field test program, the three Services [Army, Naval Operations, and Air Force] should conduct field tests on a scale sufficient to determine the military worth of candidate agent-munition combinations, their offensive uses and the means of defense against them. In addition, indoctrination in BW through the use of orientation teams, courses of instruction, on-the-job training, or by other appropriate means should be undertaken in each Service, with such mutual assistance and cooperation as may be practicable.[5]

Despite intensive development and testing, throughout this era biological weapons were neither assimilated into the thinking and planning of the regular military nor used by the United States or its partners in this endeavor—the United Kingdom and Canada and, later, Australia.

Two wars presented opportunities, the Korean War (1950–1953) and the Vietnam War (1962–1975). The Korean conflict involved the United States as the leader of a multilateral, United Nations effort that would have dampened any impulse to use biological weapons. In addition, the US program, still struggling to develop more efficient biological bombs for the dispersal of anthrax spores and other pathogens, had little reliable to offer. Indeed, in 1952 China and North Korea accused the United States of airplane attacks with contaminated insects, feathers, and other vectors, at the technical level of the Japanese program.

During the Vietnam War, the national and international controversy over US use of chemicals in combat, especially riot control agents, suggested that the introduction of biological weapons, covertly or not, would have caused a furor. During the 1960s, field tests showed increasing promise. Those involving tropical diseases and designed to penetrate jungle canopies might have tempted some military planners frustrated by Vietcong guerrilla tactics and wanting every attack option, but US government policy held the line.

Biological Warfare, the Atomic Bomb, and the Soviet Threat

In May 1946, based on a leak from a closed hearing on the Navy budget, the Associated Press carried a sensational story about a germ weapon "far more deadly than the atomic bomb" and "capable of wiping out large cities and entire crops in a single blow."[6] AP withdrew the story within hours, stating that no such weapon existed. In this and other ways, the public was prepared to perceive biological weapons as a new order of threat. The US press was then discovering the old wartime biological weapons program, with help from the Army and the Navy that emphasized the defensive measures that the US military had developed. The Navy, for example, held a press conference reporting that its research unit in Oakland, California, had used fifty San Quentin prisoners to study bubonic plague as a means of developing medical defenses. *The New York Times*, the Washington *Times-Herald, Newsweek, Life, The Nation, Science Illustrated,* and a score of other publications picked up on the postwar research publications from Camp Detrick and ran stories on the frightening strategic potential of biological weapons. A headline in *Science News Letter* read, "Germ Warfare Advances More Potent Than A-Bomb."

In 1947, as the immunity bargain with Japanese scientists was being negotiated and as the Cold War commenced, the United States submitted a draft resolution to the United Nations that defined weapons of mass destruction as including "atomic explosives, radioactive material, lethal chemical and biological weapons, and any weapons developed in the future which have characteristics comparable in destructive effect to those of the atomic bomb or other weapons mentioned above." This grouping of unproven biological weapons with chemical and nuclear weapons continued through subsequent administrations, Democratic and Republican.

With the Cold War escalating, the obvious question within government was whether the USSR posed a biological weapons threat to which the West, particularly the United States, should respond by having an offensive capability. US military and intelligence assessments acknowledged the devastating effects of the war on the Soviet economy and on health of Soviet civilians. There were known gaps between Western and Soviet technology—especially with regard to the manufacture of antibiotics and industrial fermentation capacity.[7] Yet the Soviet Union had the scientific means and the intent to pursue nuclear weapons; might it also aspire to biological weapons? Suspicions of a Soviet biological weapons program became common in Washington. In 1948, for instance, US Senator Edwin Johnson told the press that he had it on good authority that "the Russians have perfected this terrible warfare weapon of spreading plagues and germs."[8]

After the Soviet Union tested its first atomic bomb in August 1949, US military analysts speculated that biological weapons would be its next strategic priority. By 1951, although intelligence on the subject was poor, a JCS report concluded that the USSR had biological weapons and was using human subjects to test them in field trials: "Soviet interest in the BW field is of long standing. Unconfirmed reports indicate that considerable effort has been exerted and that this effort is continuing. It must be assumed, and it would be foolhardy to do otherwise, that the Soviets have developed an offensive capability in the field of biological warfare. We can be sure that there is no hesitancy on the part of the Soviets to obtain factual information on the effectiveness of their own BW agents and munitions by large-scale tests using human subjects."[9]

The Biological Weapons Startup

In 1946, the annual US Chemical Warfare Service (CWS) budget was reduced to $933,000, down from its peak of $2.4 million during the war.[10] In 1947, as the Cold War began, its allocation returned to its wartime level, $2.75 million. The postwar discovery of German nerve gas, unknown in the West, had moved the CWS in a new direction of research. Biological weapons were another reason for increased funding. Under Alden Waitt, commander of the chemical program, those in charge of biological weapons streamlined their research, concentrating on intensive studies of known pathogens and toxins and their adaptation to bombs and aerosol generators. Qualified young scientists were recruited to the program, helped by fellowships from the National Research Council of the National Academy of Sciences. Available records indicate that the postwar program went through difficult periods, due mostly to technical problems, until it reached relative stability in 1954.

The Army's postwar biological warfare research, development, and pilot-scale production program was based at Camp (later Fort) Detrick in Maryland, with additional facilities at the animal research station at Plum Island (Fort Terry), New York. Detrick also participated in projects at the Navy's Oakland Biological Laboratory in California and in various covert intelligence projects. Field testing was centered at Dugway Proving Ground in Utah. The program was tasked with procuring empty bombs from civilian contractors, filling them with biological agents or innocuous simulant payloads, and transporting them to test sites. The biological weapons program also used a variety of environmental sites in the United States and elsewhere for open-air testing, drawing on its partnership with Canada for the use of sites there and cooperating with the United Kingdom in its field tests. When the Chemical Corps requested a new facility for biological agent and munitions production to replace the wartime Vigo plant, Congress responded with funding for Plant X-201 at Pine Bluff, Arkansas. The ten-story facility, started in 1950 and completed in 1953, cost $90 million and eventually employed 1,700 workers.[11]

The United Kingdom's program at Porton Down seems to have made a smoother transition to strategic objectives and expanded field trials than the US program. David Henderson, Paul Fildes's second-in-command, took over as director, and the program maintained knowledgeable advisors, such as Paul Fildes, Lord Stamp, and Maurice Hankey, who also offered continuity. The British were against revelations about their wartime activities, which nonetheless did attract unwelcome press attention. As in the United States, the UK program was drawn to the development of biological weapons on a par with atomic ones. Having ratified the Geneva Protocol, its struggles to resist fully offensive objectives were more evident than in the United States, and at various times its program leadership claimed a purely defensive mission or appeared to limit itself to cooperation with special operations. If biological weapons had a future, though, it was for effective, large-area attacks. The postwar program at Porton Down, therefore, was committed to the same two goals as the American one: a vigorous laboratory investigation of potential agents and the initiation of large field tests with pathogens, toxins, or their simulants.[12]

In 1945, the Biology Department at Porton was renamed the Microbiology Research Department (MRD). In 1948, despite the United Kingdom's damaged economy and a shortage of construction materials, ground was broken at Porton for an immense new facility, the largest brick building in Europe at the time.[13] Henderson felt that understanding inhalatory infection should be the focus of laboratory research,[14] and he arranged for the building of a bomb-bursting chamber for expanded aerosol testing. Soon after, at Camp Detrick, the Americans built the forty-foot-high, million-liter Horton Test Sphere, the largest in the world. The Horton Sphere allowed explosive munitions to be detonated inside the chamber, with caged test animals exposed through portals.

Finding scientists to work at Porton was difficult; the responses to job advertisements were disappointingly low. Scientific advisors took a dim view of postwar media commentaries they felt turned the public and perhaps prospective staff against the program.[15] In any event, the Porton laboratories, designed for 130 scientific researchers and technicians, employed only 72. The program did, though, attract quality researchers who contributed to defenses against biological weapons. For example, biochemist Harry Smith, who arrived at Porton in 1948, discovered the three toxins produced by anthrax bacteria.[16] This discovery became the basis for the creation of the antigenic anthrax vaccine, still in use a half-century later, and for modern antitoxins.

Large-scale production was also on the Porton agenda. The MRD was funded for another large facility and two satellite pilot fermentation plants. Should a war break out with the Soviet Union, it was reckoned, Porton should be ready with the necessary production rate of thirty-five tons of agent per day, to supply the four-ton bomb capacity for each of five hundred aircraft conducting four sorties per month. Restraints from the British Cabinet and elsewhere among civilian leaders in government soon undermined this goal by imposing restrictions on building.[17]

Technical Restraints

As early as November 1945, the UK Chiefs of Staff were speculating about the future of biological weapons in ten years' time, compared to that of nuclear weapons. A War Office memo voiced optimism about biological weapons, noting that, with more support, the program might produce a bomb that could cause a 50 percent risk of death in an urban area of three square miles.[18]

In 1946, the UK Chiefs of Staff were committed to preparing the nation for the use of weapons of mass destruction, which they identified as nuclear, chemical, and biological. Under an Air Staff research project called "Red Admiral," Porton was to assist in creating an antipersonnel bomb by 1955. The Air Staff reported that despite its aversion to initiating a biological attack, as last resort capacity, it wanted a thousand-pound cluster bomb, a "parent" bomb containing many small "child" bombs, for wide-area coverage. Its estimate, following from the previous year, was that two hundred tons of this type of bomb could cause 50 percent death rates in an industrial city of one hundred square miles. High-altitude bombing was preferred, for which Porton would have to develop accurate meteorological reports and protect the disease agents from extreme cold. The report also considered spray generators, if researchers could improve their targeting.[19]

Neither the mechanical effectiveness of biological bombs in generating respirable aerosols nor the dose response for such aerosols was sufficiently tested to allow accurate estimates. Persistent technical problems had troubled the modified anthrax bomb during the late years of the war and continued to do so. For dangerous pathogens without reliable antibiotic therapy, scientists were limited to animal experiments from which they attempted to extrapolate the human dose response, that is, what inhaled dose of an agent might infect what proportion of a target population.

The US Air Force was initially enthusiastic about the potential of biological weapons, but its enthusiasm soon wore thin, as did that of the British Air Staff. By 1950, although flush with funding, biological-warfare scientists still had not perfected the four-pound Mark I prototype bomb and so decided to shift priority to a two-pound munition for cluster bombs. In February 1951, the JCS placed biological weapons in Strategic Category 1 and charged the Air Force with developing a worldwide combat and defensive capacity.[20] The biological weapons program still had not produced an efficient bomb. This failure was a sharp disappointment for the Air Force, which had placed procurement orders before waiting for bomb tests. The program's other invention, the four-pound M33 bomb, had only small area coverage, and its carriage design was already obsolete for new high-performance bombers. "Research and development had failed to support planning estimates," Dorothy Miller, an official historian of the Air Force biological program, wrote, "and the near future held out small hope for substantial improvement."[21]

The Air Force staffs involved with biological warfare were organizationally divided; one was concerned with research and development and the other with logistics and training. The first office was unsure of the technical future of biological weapons. Logistics and training had also lagged. Few military personnel knew how to handle or transport biological weapons or what to do in the case of accidents, or even how the bombs would be transported from the Chemical Corps for secure loading at airfields, or who would be in charge once the munitions went overseas. As Miller observed, "Logistical support of munitions was a nightmare."[22] A plan to train a hundred medical and science officers in biological warfare fizzled; the Chemical Corps had little to tell the Air Force that was not speculation.

Korean War Accusations

In this unsettled period, allegations of US use of biological weapons in the Korean War became an international issue, debated at the United Nations. In early 1952, the Chinese and North Korean governments accused the United States of using biological warfare tactics learned from the Japanese.

Rejecting the International Red Cross and the World Health Organization as biased toward the West, the North Koreans and the Chinese appealed to the World Peace Council to form an investigatory committee. This international committee of scientists left in June 1952 and spent two months in China and North Korea evaluating evidence for the allegation. The committee interviewed eyewitnesses and physicians, as well as four American captives (a pilot and three navigators) who had testified to US use of biological agents. The leader of the committee was the eminent British chemist and historian of Chinese technology, Joseph Needham, who had been stationed in China during World War II. Soviet biologist N. N. Zhukov-Verezhnikov, who had been the chief medical expert at the 1949 Khabarovsk trial, was also a member. Although it conducted no research of its own, the committee concluded that the United States had conducted biological warfare, primarily by airdrops of agents or contaminated materials. It compiled a seven-hundred-page report detailing fifty alleged incidents.[23] The report included autopsy reports and testimony by physicians who had treated patients and scientists who had analyzed data, along with photostatic copies of supporting statements handwritten by the captured US airmen. The Needham report, as it is often called, contains allusions to the Japanese use of disease agents in China and includes claims that the US government was continuing to protect Japan from its criminal past. The report also cited postwar published articles by US biological warfare scientists as proof of an American offensive program and reminded readers that the United States had failed to ratify the Geneva Protocol.

The US government denied the charges of biological weapons use, and this denial was generally supported in the West and by eminent scientists who challenged the Needham report. The allegations are still a matter for study. Recently, for example, a Japanese journalist claimed to have seen Soviet documents indicating that the evidence of the Korean War germ attacks had been faked.[24] The controversy demonstrated the difficulty of separating evidence from ideology in the case of suspicious disease outbreaks and also reinforced the dictum that intelligence in time of war is at best faulty. With ideological adversaries locked in battle, no credible independent investigation, allowing scientists to pose critical questions and collect their own data, seemed possible.

Toward Improved Testing

Meanwhile, within the secret US program, scientists were having problems producing evidence that their biological weapons could be effective. Either the biological bomb tests were inconclusive or their more promising outcomes were poorly communicated to Air Force personnel, who remained unpersuaded. Specifically, Air Force officials had grown impatient with the biological program's reliance on animal data. They wanted medical accuracy that would allow better bombs. Yet human-subjects research using biological agents could only be conducted if quick and reliable postexposure therapies were available, or else the research was ethically impossible.

In April 1953, recommendations to cancel all procurement agreements with the biological weapons program were circulating through the Air Force, at the level of the Secretary. One memo referred to the program as "an unwarranted luxury."[25] Rather than gaining confidence in biological weapons, Air Force staff perceived the program as an unstable environment, with support liable to "booms and busts." Compared with the larger conventional and nuclear programs, it had limited career value for talented officers.[26]

Air Force confidence in biological weapons was at an ebb, but the Army continued its support. On 5 March 1954, a new directive from the Defense Department limited the program to long-range research and development, coupled with field testing. The biological program would research only improved munitions and concentrate on "lethal anti-personnel biological munitions for release from high-speed aircraft [the one-half pound E61 anthrax bomb] and upon the development of munitions for delivery of anti-crop pathogens from high speed aircraft and by balloons."[27] An "acute disillusionment" experienced by former enthusiasts translated into budget cuts.[28] The Air Force investment in biological weapons research and development, at a high of $5.7 million in 1953 fell to $1 million in 1956. Other Air Force categories, including the purchase of antipersonnel munitions from CWS and industry, showed similar declines.

The biological weapons program had two ways of increasing confidence in its weapons, and its scientists used both of them. The first effort was to improve field tests, which included secret sea and land trials with real pathogens and with simulants over UK, US, and Canadian urban targets. Improved estimates of aerosol dispersion patterns and agent survivability and virulence in various settings, under various weather conditions, relied on these data. As they grew larger, the field trials also were a way of gaining operational experience relevant to the conduct of strategic biological attack, uncoupled from the real-world contingencies of causing disease outbreaks and the radical consequences for war and society.[29]

The second way to improve confidence in estimates of biological weapons effects was to embark on human-subjects research with pathogenic aerosols. After the war, these experiments were made possible through the availability of antibiotics for postexposure treatment, particularly for curing tularemia. The calculation of dose response was critical. It had to be known at least approximately in order to establish requirements for munitions performance and to estimate how many munitions would be required to attack any particular target. For its part, the Air Force needed to know how many bombs would create a desired target effect so it could plan how many planes would be needed. The UK and US open-air field trials with simulants and with real agents were tests of the feasibility of this grand plan.

Operation Harness

As with much biological warfare research, the British were first with both sea trials and the investigation of the dispersal of biological agents over urban areas. "Operation Harness," the 1948–1949 British series of field trials off the Caribbean islands of Antigua and St. Kitts, was the first in a limited series sponsored by the United Kingdom. The United States followed suit and continued sea trials for years.[30] The choice to hold the trials at sea was determined by fears of soil contamination (especially with anthrax, which had badly contaminated Gruinard Island), of public disclosure, and of possible danger to local populations. Unfortunately, high winds in the Caribbean forced the testing closer to land, where they were conducted "with many small fishing craft in the vicinity."[31]

American scientists cooperated in Operation Harness, supplying some pathogens and bombs (the four-pound American E48.R2 type), along with bursters and modified fuses. The bombs were filled onsite and then suspended fuse downward for detonation twelve to eighteen inches above the ocean surface, for dispersal over caged animals on dinghies. The test animals included guinea pigs, sheep, and monkeys imported from India; it was the first time that guinea pigs and monkeys were used in anthrax aerosol tests, and both would frequently be used in subsequent research. Over the course of about two months, twenty-seven trials with bacteria, including some dispersed by spray generators, were carried out. David Henderson later reported that major miscalculations had been made, that the project had been rushed and its results were few. Although he believed the sea trials could be improved if plans were better, he also argued for better scientific investigation of aerosols, even referring to field trials as a "blunderbuss method" to gain knowledge about dispersal processes.[32] The summary judgment from UK authorities was that more such trials should be conducted, and some were. It was the United States, though, that embarked on dozens more such tests of increasing sophistication and elaboration, and that turned its attention to experimental aerial tests in urban areas, as the British had also tried, and from there to large regional simulations.

The St Jo Program

In the United States, the Air Force wanted "a methodology for predicting weapons effects and for estimating munitions expenditures. To acquire this methodology it needed to understand the predictable dispersal of aerosol clouds over the potential target areas."[33] The request was reasonable. By 1951, the Chemical Corps intended to give the investigation of aerosols top priority, and expenditures of over $4 million were expected for the period 1950–54. The technical goal was uniform dissemination of agents of around one to seven microns in particle size, the optimal size range for human inhalation, which could lead, for example, to anthrax infection. The postwar program had kept its ties to industry; several types of spray generators were being developed by independent contractors.

In response to the Air Force demand, Detrick scientists put together the St Jo Munitions Expenditure Panel. Its mission, begun in 1953, was to simulate anthrax attacks on urban targets. Three North American cities were chosen to approximate Soviet cities: St. Louis, Minneapolis, and Winnipeg.[34] From January to September, its experimenters used generators mounted atop automobiles parked in various urban locations to disperse clouds of simulants. The presumed conditions of the potential attack were stated clearly: "Any expenditure figures finally derived will refer to a completely unprotected target population, which is assumed to be exposed in the open in a city, during the whole time of passage of the biological cloud."[35]

In the late 1940s, the British had already tested an urban target by dispersing simulants on the city of Salisbury, near Porton, after which they carefully measured the influence of meteorology on cloud dispersion.[36] In 1950, US ships had secretly sprayed aerosols of the anthrax simulant *Bacillus globigii* (BG) and *Serratia marcescens* (the tracking germ referred to in the 1934 Wickham Steed report) to be carried by wind over the San Francisco Bay Area.[37]

The St Jo program brought a new array of science to bear on offensive simulations. Its first trials, which ran intermittently from January until September 1953, integrated every data source Detrick experts had: munitions testing and animal dose-response studies at Detrick, open-air trials at Dugway using BG, and meteorological studies commissioned jointly from Stanford University experts and a private contractor.[38] The scientists in charge used data from the UK Salisbury tests and from earlier CWS experiments with chemical dispersion. The program also enlisted the help of computer programmers from the National Bureau of Standards.

In August 1954, the St Jo panel published 101 pages of methodology, with maps of aerosol dispersal patterns from urban trials and dozens of closely argued mathematically formulas, concerning predicted doses of anthrax and the expenditure of bombs it would take to produce them.[39] From the Air Force perspective, the St Jo program "represented a major advance in the concept for testing a biological munition-agent combination."[40] Most important, it showed how, depending on meteorology and terrain, the number and pattern of bombs dropped on a target were related to the percentage of "casualty production" from anthrax aerosols.

The panel for the St Jo program tempered its report with scientific skepticism and caveats. For example, it reckoned that eight thousand anthrax spores would infect 50 percent of humans exposed (the lethal dose for half the population, or LD50), but cautioned that this figure was only an extrapolation from monkey research data. It noted that the E61R4 bomb had limited aerosol efficiency. An effective aerosol would deliver anthrax particles in the one- to seven-micron range; only 2 percent of the aerosol from the E61R4 was in this range. Adding more bombs or increasing the virulence of the agent (which had been done for anthrax) could increase casualties.

St Jo gave needed new hope to the biological weapons program by pointing to the potential of aerosols and bombs, if only the technical obstacles could be overcome. The Air Force consequently became interested in those technicalities—what the program could do, for example, to increase an agent's virulence or improve aerosols by using dry agents instead of slurries. In another development, it asked for more study of less lethal "incapacitating" diseases, such as brucellosis, and of highly contagious diseases, including smallpox.[41]

Field trials remained essential to the program, and their numbers escalated. From 1950 to 1954, seventy-nine trials of BG were staged, including sea trials off the coasts of Virginia, California, and Florida. BG land trials took place at Dugway Proving Ground; Camp Detrick; Elgin Air Force Base; Fort McClellan, Alabama; Fort Belvoir, Virginia; Wright-Patterson Air Force Base; Loring Air Force Base, Maine; and Kirtland Air Force Base, New Mexico. In 1955, two dispersion tests were conducted on major state highways in Pennsylvania. From 1951 to 1955, at Dugway, there were twenty-five open-air field trials with pathogenic agents, including those for anthrax, plague, brucellosis, Q fever, and tularemia. Some of these trials, for example, a 1954 *B. anthracis* trial, ran for months. The use of fluorescent particles to test wind dispersal in public places increased greatly once the Stanford University meteorological test results were reported. In addition to the tests in Minneapolis and St. Louis for the St Jo program and the Bay Area tests, open-air trials were conducted at Corpus Christi, Texas; in the Mojave Desert; and, in 1957 and 1958, in unspecified areas east of the Rocky Mountains. In 1959 and 1960, thirteen trials were conducted in unspecified locations in north central Texas.[42] This activity would increase sharply as the program entered the 1960s.

Project Whitecoat

A key phrase in the St Jo methodology report was the disclaimer about the lethal dose of anthrax in humans, that the dose response could not be proven without actually exposing human subjects to anthrax aerosol, which would be highly dangerous.[43] Because, unlike anthrax, the bacterial agent of tularemia (*Francisella tularensis*) could be rapidly and reliably cured by early administration of antibiotics, it held promise for human research. Although not as lethal as inhalational anthrax, tularemia (also known as rabbit fever) could cause debilitating disease and death rates of as high as 50 percent, depending on the strain.

In 1952, at the time when the United States suspected the Soviets of conducting unethical field trials using human subjects, physicians delegated to Camp Detrick by the Surgeon General's Office had the idea to start tularemia research. Their first attempt to use enlisted men at the base caused a sit-down strike. Then they considered research on Seventh Day Adventists who were conscientious objectors. Starting in 1953, after an informal agreement with leaders of the Seventh Day Adventist (SDA) Church, the medical corps began a long series of weapons tests on church volunteers. The pilot phase was called CD-22 and involved ninety-one volunteers' being exposed to aerosols of Q fever, which could also be reliably cured by antibiotics. In 1954, on the basis of the success of CD-22, Camp Detrick moved forward with Project Whitecoat.[44]

Twice a year a US Army officer and a church elder traveled to Fort Sam Houston, Texas, where conscientious objectors were trained as medics. There they recruited only Seventh Day Adventists for the Whitecoat project. Around 2,200 SDA men volunteered in the course of the program. From 1955 to 1957, trials were based on injections of *Pasteurella tularensis.* In September 1957, a project officer began a series of "respiratory challenges," that is, aerosol exposure in order to gauge dose response. Some volunteers were sick for days or weeks despite antibiotic therapy. Others recovered quickly. Tests were also conducted on how, once infected at various doses, the men were able to perform simple tasks or do mathematical calculations.

Nearly two hundred human-subjects experiments were conducted under Project Whitecoat. The peak years were 1963 through 1966, when most of the experiments involved tularemia, Venezuelan equine encephalitis, or cutaneous anthrax, a form for which antibiotics were fully effective, in contrast to anthrax acquired by inhalation. In 1967 through 1969, as the program was ending, a switch was made to studying tropical diseases, primarily sand fly fever and yellow fever, but also Rift Valley fever, plague, and Chikungunya, a mosquito-borne virus. Project Whitecoat ended with the 1973 abolishment of the draft, although Army history shows that some tests continued until 1976.

The Project Whitecoat participants were physically well treated and seemed, from Army checkups over the years, to suffer no long-term health consequences. The problematic issue was their ignorance of the goal of the research. Church leaders had understood that the project was for defensive purposes, to discover maximum protection against tularemia. Yet the main goal of the project was offensive, to ascertain human dose response for tularemia and thereby further the development of an effective bomb and, from there, the military assimilation of biological weapons for strategic attacks.

Industry was another source of human data on biological weapons. In the 1950s, in order to conduct control studies on its anthrax vaccine, the Army sent scientists to woolen and textile mills and other factory settings known for high rates of anthrax spore contamination. In 1957, in a felt-making factory in Manchester, New Hampshire, five unvaccinated workers came down with inhalation anthrax, probably from a highly contaminated shipment of imported animal hair or wool, and four died.[45] The investigatory work on those deaths, including autopsies, added to the Japanese data. As occupational health standards improved and as the textile industries grew outmoded or went overseas, the factory environment for studying anthrax disappeared.

Options for First Strike

As the biological weapons program struggled for stability, a major policy change took place that would free the biological and chemical programs from the constraints implied by earlier US tacit acceptance of the Geneva Protocol, short of ratification, and bring decision making for both in line with nuclear weapons policy. At the time, US nuclear policy left the ultimate decision for use with the president but allowed considerable latitude for the Defense Department to plan for possible first strikes. Nuclear deterrence played a part in this policy. Deterrence had also played a part in restraining the use of chemical weapons in World War II. The role of deterrence regarding biological weapons was much more ambiguous. Rather than deter the Soviet Union by threatening it with disease weapons, the United States was proceeding in secret to develop a new strategic weapon. Advocates of this venture within the military and the few members of Congress who approved the CWS budget wanted no restraints on the program.

In 1950, US policy for biological weapons was retaliation-only, in line with the Geneva Protocol. This policy was described as "interim," meaning that it could change as weapons were improved and reevaluated.[46] In May 1954, the Army Chief of Staff called for a review of retaliation-only use of biological and chemical weapons. As Miller recorded, the argument about this restraint was longstanding. The Army felt that "a major obstacle to BW progress was the reluctance of the military to invest time and resources in a weapons system that might never be used."[47] With a first-use option, a full development and military assimilation could take place: the United States could pick the time, place, and terms of an attack instead of merely reacting. And it would not be limited to attacks against the Soviet Union; it could attack any enemy. Critics within the military argued that germ weapons were simply not competitive with conventional and nuclear weapons and that only the requirement for a retaliatory capability against the Soviet Union kept the program alive.[48]

The decision about a policy change was ultimately passed up to high-level security advisors who decided to free the Army from the retaliation-only use of both biological and chemical weapons. The justification cited was a new intimation of a Soviet biological threat. The source of concern was the February 20 speech of Soviet Marshall Georgiy Zhukov, the same military leader whose troops had been the target of Japanese biological weapons in 1939. At the twentieth Communist Party of the Soviet Union Congress, Zhukov declared, "Future war, if they unleash it, will be characterized by the massive use of air forces, various rocket weapons, and various means of mass destruction, such as atomic, thermonuclear, chemical and bacteriological weapons."[49] Security advisors took this statement as evidence of Soviet biological weapons capacity and its intention to use them for mass destruction in future wars.

By 1956, a revised biological and chemical weapons policy was formulated to the effect that the United States would be prepared to use such weapons in a general war to enhance military effectiveness. The decision to use them would be reserved to the president.[50] That year, the US Army manual *The Law of Land Warfare* deleted reference to "retaliation only" and asserted instead, "The United States is not a party to any treaty, now in force, that prohibits or restricts the use in warfare of toxic or nontoxic gases, or smoke or incendiary materials or of bacteriological warfare."[51] In the US Army manual that articulated doctrine for all the armed forces, reliance on presidential authority was clearly expressed, leaving latitude for offensive development as in the nuclear program: "The decision for U.S. forces to use chemical and biological weapons rests with the President of the United States."[52]

A major effect of this policy change was the expansion of plans for what biological weapons might accomplish. Large-scale simulations took on more importance. At Dugway, $2.8 million was invested in an improved grid system and meteorological network to accommodate large-scale aerial bombing and spray tests. In 1955, at the tripartite meetings with the United Kingdom and Canada, the US government announced its new approach, the "Large Area Concept" (LAC), whereby an airplane or a ship could spread an off-target line of pathogens and possibly infect an entire region.[53] One major supposition was that meteorological conditions could be predicted for attack purposes, and the other was that pathogens could survive for long periods while airborne.

In 1957, to test the first supposition, UK scientists embarked on field trials in which fluorescent zinc cadmium sulfide particles were released from an aircraft flying upwind of a target area. The first tests involved releasing particles along a three-hundred-mile line over the North Sea and then taking air samples in England and Wales to track their dispersal. It was found that millions of people could be exposed to a biological aerosol. In 1957, as part of Operation LAC, the US military also disseminated particles of zinc cadmium sulfide along a path from South Dakota to Minnesota. A repeat trial in February 1958 at Dugway Proving Ground showed that six-hundred-mile dispersal could affect millions. Further tests with live simulants suggested that some pathogens could, in fact, survive these tests.

Expansion in the Kennedy-Johnson Years

The extent to which any head of state is closely advised of biological weapons policy concerns historians of every program. The US postwar chemical and biological weapons programs were small enough, within the context of the Defense Department, to escape attention from high civilian officials. In a 1960 press conference, President Eisenhower was asked whether a policy change toward first use had occurred. From the military perspective, the change had taken place. Yet Eisenhower replied that no such official suggestion had been made to him, and he added, "so far as my own instinct is concerned it is to not start such a thing as that first."[54]

Shortly after President John F. Kennedy's 1961 inauguration, Secretary of Defense Robert McNamara began organizational changes to centralize and streamline Pentagon authority. One result was that the biological weapons program achieved more autonomy from the Chemical Corps. It was placed under the newly formed Munitions Command, within Army Materiel Command. The renamed Fort Detrick, its center, maintained operational control over the Pine Bluff plant. Only the Army had the right to work on highly infectious agents, while the Air Force and Navy kept their research and procurement programs, estimated to account for 35 percent of overall chemical and biological spending, which in the next several years reached an annual $300 million. Defense also instituted a Joint Technical Coordinating Committee to review relevant programs and coordinate research, but none of this activity constituted a high-level review.[55]

The 1960s accelerated the pace of biological-weapons field testing. The escalation of the Vietnam War increased military authority over weapons development, and the CWS benefited from this expanded latitude. This period saw a great increase in the variety and scale of simulated civilian targeting, reinforced by improved information on dose response from human subjects research and by other research and technical improvements that came from numerous university and privately contracted projects. From November 1962 to April 1967, for example, Dugway Proving Grounds was nearly constantly involved in large-scale field tests of *P. tularensis.* Some tests had defensive purposes as well. In 1965, BG spores were dispersed at National Airport and Greyhound Terminal in Washington, DC; in the New York City subway in 1966; and off the coast of California, near San Clemente, where President Richard Nixon had his West Coast retreat, in August through September 1968. In 1964–65, simulated "anti-animal" attacks were staged at six major stockyards in Texas, Missouri, Minnesota, Iowa, Nebraska, and South Dakota. Most tests, however, were to demonstrate regional attack scenarios. Between June 1961 and November 1969, twenty-two fluorescent particle trials were staged in public domain areas, such as national forests in South Carolina and Minnesota, and in Fort Wayne, Indiana; Corpus Christi, Texas; and, again in 1968, San Francisco. St. Louis was retested three more times, from May to September 1963, April to October 1964, and March 1965, in a recapitulation of the seasonal experiments of the St Jo project.

An important innovation within the biological program was Project 112, a land and sea project for expanded testing of chemical and biological weapons. In 1962, the Army established the new Deseret Test Center, at Fort Douglas, Utah, from which the project was coordinated. The Army, Navy, and Air Force were involved; Canada and the United Kingdom were partners. Although two tests in 1968 simulated anthrax attacks on warships and how the crews might be protected, the majority of Project 112 trials appears to have been for testing the offensive capacity of biological weapons.

At least fifty Project 112 trials took place, most of them from December 1962 to 1970. Public information remains incomplete on these and the related sea trials for which the backdrop was the escalating Vietnam War. Eighteen of these trials involved simulants of biological weapons (usually BG), nearly all in the 1963–65 period. The years 1965 to 1967 were peak ones for chemical weapons tests, with a total of fourteen trials of VX, sarin, nerve gas simulants, and tear gases. In 1967 and 1968, tests of chemical agents were staged at Porton Down and at Ralston, Canada. Otherwise, biological and chemical weapons tests were often conducted at the same sites or regions, at different times.

The sea trials conducted under Project 112 were collectively called Shipboard Hazard and Defense (SHAD). SHAD involved at least thirteen warships, along with bombers and airplanes fitted with spray generators. One sea trial, named "Big Tom," was the epitome of what the Army intended with Project 112. Held off the coast of Oahu, Hawaii in May–June 1965, it brought together the Navy and Air Force to "evaluate the feasibility of a biological attack against an island complex and to evaluate doctrine and tactics for delivery of such an attack."[56] The test consisted of a series of off-target releases of both liquid BG (from a spray tank mounted on a Navy A-4 aircraft) and dry BG from a spray tank mounted on a USAF F-105 aircraft. The experimenters were concerned with area coverage and in particular with how much aerosol would penetrate the jungle canopy. In 1964, another trial, called Shady Grove, was conducted near Johnston Atoll, eight hundred miles west of Hawaii, where the US military also stored chemical weapons. Shady Grove tested actual pathogens, for tularemia and Q fever, on rhesus monkeys in cages on the decks of ships.

The development of offensive strategies for the Vietnam War was a likely goal for some of the tests. For instance, the trial "Yellow Leaf" was to test competing Army and Navy biological weapons systems against targets in a jungle environment. The trial began in February 1964 with a series of 184 bomb tests at the Fort Sherman Military Reservation in Panama. Political upheaval required moving the tests to the Olaa Forest in Hawaii, where another hundred tests were staged. Also at the Olaa Forest, the Army conducted twenty aerosol tests to gauge diffusion beneath a jungle canopy. In 1965, another trial, called "Magic Sword," was conducted to see if mosquitoes released in a Pacific Island setting might convey infectious diseases such as dengue fever or yellow fever.

Other field tests took place near Fort Greely, Alaska, to gauge biological warfare under USSR winter conditions. One test called "Red Cloud" was conducted from late November 1966 to mid-February 1967, to evaluate the dissemination of wet and dry forms of the agent of tularemia in extremely frigid conditions. The tests expanded to the Pacific, with a variety of simulants and pathogens, to show that the program was ready for war.

In September and October of 1968, the final and probably most elaborate trial (DTC Test 68–50) took place at the Eniwetok Atoll in the Marshall Islands.[57] Staphylococcal enterotoxin, Type B (SEB, an incapacitating toxin produced by a common bacterium, *Staphylococcus aureus*) and BG were dispersed by F4 Phantom jets with newly designed dissemination tanks. The conclusion from this trial was that a single tank could disperse a virulent infectious agent over nearly a thousand square miles.

With every test, the US military increased its readiness for strategic strikes with biological agents. During the 1960s, unfettered by any retaliation-only policy, the Army showed that it had met the challenge of large-scale attack, at least by its own calculations. It appeared only to await a command from a higher authority. That command never came. In 1969, the program's enormous effort to show that biological warfare could rival nuclear warfare—the laboratory and human subjects research, the industrial production and stockpiling of agents, the manufacture of bombs and spray generators, fitting of airplanes and ships for dispersal, the indoctrination of troops, and the large-scale field trials—would end abruptly with a brief presidential statement about the danger these weapons posed to humanity.

CHAPTER 6

The Nixon Decision

While US field tests of biological weapons were reaching new heights of realism, political forces were at play that would make all offensive biological weapons programs illegal. The era of legitimate state retaliation in kind, begun by the French in the 1920s, ended with President Nixon's 1969 renunciation of biological weapons for the US and with the 1972 Biological Weapons Convention. By 1975, when the treaty came into force and also when the US finally ratified the Geneva Protocol, comprehensive international legal norms against state programs and the possession or use of biological weapons were in place.

In the United Kingdom this shift away from biological warfare programs began earlier, in 1959, when Porton declared an end to its offensive biological projects in favor of defensive research and development. Having ratified the Geneva Protocol in 1930 and reserved only the right of reprisal, the United Kingdom's adherence to the treaty was an obstacle to integrating either biological or chemical weapons into war plans: why develop weapons that might never be used?[1] Compared to nuclear weapons, biological weapons offered little additional as a deterrent. Technically they were untested in war and, despite large-scale simulations, it remained that the likely effects of actual intentional epidemics would be delayed, uncertain, and geographically imprecise. In 1952 the British successfully tested their first atomic bomb, giving the United Kingdom a new level of strategic and deterrent capability. The French, having conducted their first atomic bomb test in the Sahara in 1960, were on the same road toward becoming a nuclear power, and their interest in biological warfare was fast waning.

These retreats left the United States and the Soviet Union as the two major states with the intent and capability to wage war with unconventional weapons. During the 1960s, Western European nations became increasingly concerned about becoming the battleground where the two superpowers might resort to nuclear, chemical, or biological weapons.

Military Latitude

The US military's defiance of the Geneva Protocol restraints was demonstrated during the Vietnam War, when riot-control agents were intensively used as battlefield weapons. Beforehand, during the Eisenhower administration, military advocates argued for the latitude to use chemical and biological agents categorized as "incapacitating." At a February 1960 meeting of the National Security Council, President Eisenhower and members of his cabinet were briefed by the Defense Department on the merits of tear gas and biological agents of low lethality, such as Venezuelan equine encephalitis, Q fever, and Rift Valley fever.[2] Using intelligence data, the Pentagon characterized the Soviet threat as follows: a stockpile of lethal chemical and biological weapons 75 percent greater than that of the US and with thousands more trained chemical and biological warfare troops. At the February briefing, three hypothetical scenarios were presented of Communist takeovers of areas in which enemy soldiers and civilians were mixed or civilians were held hostage; while the details were vague, the scenarios suggested settings in nonindustrial countries, under conditions of guerrilla warfare. The military solution for each dilemma was large-scale "nonlethal" agent dispersal. Even though the Geneva Conventions of 1949 expressly forbade attacks against civilians, Defense officials argued that incapacitating gas and disease weapons in a time of war might be protective of civilians already likely to be killed.

The summary of this briefing indicates that President Eisenhower thought the use of such incapacitating agents was "a splendid idea," but stressed the need for adequate troop defense against the lethal chemical and biological weapons that the enemy (meaning the USSR and its allies) would probably use in retaliation. It fell to Allen Dulles, director of the Central Intelligence Agency, to make the point that biological weapons were not acceptable to the world at large, whereas tear gas was widely used in police action and therefore was the incapacitating agent that could be used "respectably."[3] This evaluation held into the Kennedy and Johnson administrations, when the United States undertook first use of the chemical riot-control agent CS in

Vietnam while refraining from any use of the biological weapons that it had tested in open-air trials.

As presidential candidates in 1960, John Kennedy and Richard Nixon were asked by the Society for Social Responsibility in Science, an early critic of the Army's programs, to state a position on chemical and biological warfare. Kennedy replied that he would look for international solutions for eventual disarmament; Nixon answered that Congress and the Defense Department were performing well in defending the nation.[4]

Nevertheless, soon after his election, President Kennedy defined a military policy based on "limited and flexible response," and this policy ultimately supported the Pentagon's prerogative to plan for first use of chemical and biological weapons.[5] In 1961, in his first address to Congress, Kennedy called for innovation in weapons development to reduce reliance on nuclear weapons, and he cited the threat of Castro's Cuba as an example of a terrain in which the United States might be involved in guerrilla warfare.[6] The Chemical Corps was poised to respond with a diverse chemical arsenal, from anti-crop agents to the mind-altering incapacitant BZ, and it was also developing and testing a diversity of biological weapons.

That same year, 1961, the Kennedy administration authorized herbicide use for defoliation and for crop destruction in Vietnam. The next year, during the Cuban missile crisis, planes were fitted with spray tanks, apparently in preparation for a Venezuelan equine encephalitis agent attack that was never authorized. In 1963, American military advisors requested and received permission for more extensive herbicide use. They also queried whether the use of riot-control agents in Vietnam would be acceptable to the US government. The State Department response was affirmative, if such agents were limited to quelling civil disturbances and similar purposes.[7]

Simultaneously, the Kennedy administration's new Arms Control and Disarmament Agency (ACDA), which reported directly to the president, began turning its attention to treaty restraints on chemical and biological weapons. By March 1962, ACDA was preparing papers proposing two international expert commissions, one for the inspection of nuclear stockpiles and the other for inspecting chemical and biological weapons stockpiles.[8]

In 1961, at the joint initiative of the United States and the Soviet Union, an Eighteen Nations Disarmament Committee (ENDC) was created. Chaired by the two powers and including nonaligned nations, it was charged with the conduct of negotiations toward an arms-control agreement to reduce Cold War dangers. Disagreement continued within the US government over disarmament goals in general and over international monitoring of treaty compliance.[9] The Joint Chiefs of Staff (JCS) was reluctant to have any treaty negotiation refer broadly to weapons of mass destruction lest US options for initiating the tactical use of chemical and biological weapons be as narrowly limited as their nuclear ones.[10]

In October 1963, ACDA director William Foster recommended to the White House a thorough interagency biological weapons policy review. On November 5, national security advisor McGeorge Bundy gave the go-ahead, and on November 12, Foster circulated a proposal for an interagency working group on national policy related to chemical and biological arms control and disarmament.[11] Ten days later, before action could be taken, President Kennedy was assassinated, and Lyndon Johnson was sworn in as his successor. The ACDA proposal was sidelined. Under Johnson, the Chemical Corps maintained and strengthened its offensive objectives, pursuing Project 112 and other large-scale attack simulations.

Herbicides and Riot-Control Agents

In 1965, a *New York Times* story filed from Saigon revealed that the US use of herbicides had escalated to broad area destruction of vegetation; American airplanes were causing widespread damage to forests and farms.[12] The two stated goals of US herbicide spraying in Vietnam were the elimination of dense foliage from strategic roads and enemy staging areas and the destruction of crops in enemy food-production areas. Chemical herbicides and insecticides were in everyday use in the United States and, with the 1962 publication of Rachel Carson's *Silent Spring,* arguments about their toxic effect on human health and the environment—as carcinogens, as the cause of chromosomal damage, as water pollutants, and as destructive of plant and animal species—were gospel to the new environmental movement.[13] The use of anticrop herbicides also raised the issue of total war doctrine: what was the difference between starving enemy civilians and bombing them? Harvard University nutritionist Jean Mayer, later a Nixon advisor, pointed out that in war the least empowered civilians—the elderly, women, and children—go without food, while soldiers are fed.[14] In addition, Vietnam had a peasant economy with limited means of storing or transporting food.

The rapidly escalating use of the tear gas CS also proved controversial. In 1964 and 1965, the US military assisted the South Vietnamese in using riot-control agents in civilian settings. The Department of Defense argued that assisting South Vietnam authorities to quell civic unrest with tear gas was no different from its legitimate use elsewhere in similar settings for the same purpose. The gases used were nonlethal, commercially available, and commonly used by police forces around the world. Secretary of State Dean Rusk argued that such use was not prohibited by the Geneva Protocol, saying, "We do not expect that gas will be used in ordinary military operations."[15]

As during the 1960 meeting attended by President Eisenhower, the Pentagon argued that its riot-control gases were being used for humane purposes, to rescue civilian hostages. To support this argument, it publicized an incident in which an American colonel reportedly used CS grenades to flush rebels out of a cave where they were holding civilian hostages. It became debatable whether such an incident had ever taken place.[16] In any case, and despite Rusk's assurances, CS soon came into massive use, and by the end of the war ten million pounds of it had been used in combat.[17]

In early 1966, the *New York Times* reported that the Army had invented a rudimentary chemical bomb, a fifty-five-gallon drum fitted with a detonator and filled with CS or other tear gases or agents that induced nausea and vomiting.[18] This drum was dropped from helicopters and planes as a preliminary to B-52 bombing raids, aircraft strikes using napalm, and artillery bombardments, which were then followed up by infantry assaults.[19] This innovation was followed by the use of some two dozen other types of CS munitions for air and ground delivery. As might have been predicted, enemy soldiers soon developed protective tactics, including the use of gas masks, and acquired US CS munitions, which they used to harass American and South Vietnamese troops.

International criticism of the United States for using chemicals in Vietnam was intense. Despite US arguments that the terms of the Geneva Protocol did not apply, the majority of nations engaged in the controversy and eminent US legal authorities argued otherwise.[20]

After World War I and the widespread acceptance of the 1925 Geneva Protocol, use of chemical agents was rare and, when it occurred, industrialized nations used it against unprotected weaker states or subgroups that could not retaliate. The Italians under Mussolini used tear gas and mustard gas in Ethiopia in 1935–36; Japan reportedly used gas against China in 1938 and later. In February 1967, as news that the Americans were using chemical weapons in Vietnam was being reported, the International Red Cross verified that Egypt had been using chemical agents against Yemeni royalists.[21] The Americans' use of chemicals in Vietnam seemed to fit this same asymmetrical pattern, in which the enemy had no defenses or retaliatory means.

Not unexpectedly, the Soviet Union and its allies condemned the United States, while America's major European allies grew troubled that the US violation of the chemical weapons norm might destabilize international efforts to prevent nuclear weapons proliferation. Even with the 1965 assurances of

Secretary Rusk that tear gas would not be used, the US use of chemicals in war also raised questions about whether the mission of winning the war in Vietnam had undermined the governing authority of President Johnson.

Scientists, the Media, and Congress

A combination of public and political pressures eventually curtailed the Pentagon's latitude in developing its chemical and biological programs. In the public sphere, scientists, journalists, and members of Congress defined the salient issues. Scientists, from microbiologists to chemists to experts in tropical plants and disease, developed professional networks with authoritative claims to interpreting the technical and political issues in this arms-control area.[22]

The collective base for science advocacy against biological weapons was well established during the postwar years, either jointly with chemical weapons or as a unique threat.[23] The Federation of American Scientists (FAS), founded in 1946, was an enduring critic of chemical and biological weapons programs and a promoter of US ratification of the Geneva Protocol. During the controversy surrounding alleged US use of biological weapons during the Korean War, its members urged the United States to clarify its policy and assure the world that the United States had "no intention of introducing biological weapons in the world's already terrifying arsenal."[24]

Pugwash, the international organization started in 1957 to bring Western and Soviet scientists together to discuss means of reducing the nuclear threat, later sought to promote treaty restraints on chemical and biological weapons.[25] In 1959, it devoted its fifth conference to the prevention of chemical and biological warfare.[26] The conference looked beyond the Geneva Protocol to a more comprehensive treaty to address weapons development and production. Compliance with such a treaty, Pugwash scientists agreed, would be difficult to insure, unless nations renounced "the miasma of secrecy" around biological and chemical weapons and worked for "the benevolent application of microbiological and chemical knowledge."[27] In 1966 the organization staged laboratory inspection exercises in four politically diverse European countries (Austria, Czechoslovakia, Denmark, and Sweden), perhaps the first verification exercises of their kind.[28] That same year, the newly formed Stockholm International Peace Research Institute (SIPRI) turned its attention to chemical and biological weapons. [29] It eventually published the canonical six-volume series *The Problems of Chemical and Biological Warfare*[30] and other related books.

During the 1960s, a half-dozen other scientific organizations, including the Union of Concerned Scientists and the Physicians for Social Responsibility, protested the US use of chemicals in Vietnam. Individual well-known scientists also protested the use of chemicals in Vietnam; their advice was not well received by the Johnson administration. In September 1966, Arthur Galston from Yale University, along with eleven other eminent plant biologists, sent a letter to the president warning about the ecological consequences of herbicide use in Vietnam. The letter did not presume to take a stand on the war itself. The State Department sent back a curt response denying that any harm was being done to the environment or civilians, who, it was asserted, were being resettled as necessary.[31] An impartial onsite investigation of the environmental consequences of herbicides was repeatedly blocked by the Pentagon. Nor was there universal consensus among scientists as to government policy. Within the American Association for the Advancement of Science (AAAS), which included many scientists from government and industry, the motion for an independent onsite inquiry was twice defeated.

Undeterred, figures such as Jean Mayer and public health physician Victor Sidel actively argued for restraints at professional meetings and congressional hearings.[32] Theodor Rosebury, coauthor of the 1942 Rosebury and Kabat report and a former division head at Camp Detrick, spoke publicly about the dangers of biological weapons to civilians. His 1949 book *Peace or Pestilence?* had predicted the problems of military secrecy and government pursuit of "public health in reverse" that came to the fore in the 1960s.[33]

In 1966, Harvard biologist Matthew Meselson, a consultant to ACDA in the Kennedy administration, and his Harvard colleague Dr. John Edsall, with assistance from the FAS, circulated a petition to President Johnson seeking review of US policies on chemical and biological weapons.[34] The petition was signed by five thousand scientists around the country, including seventeen Nobel Prize laureates and one hundred and twenty-seven members of the National Academy of Sciences.[35] Like Galston's letter, the petition voiced no opinion on the Vietnam War. Instead, it advised three steps: a comprehensive top-level review of relevant US weapons policy, a halt to the use of antipersonnel and anticrop chemical weapons in Vietnam, and a ban on first use of chemical and biological weapons. Having advance notice of the petition, Donald Hornig, President Johnson's science advisor, recommended that the president make a public statement clarifying US policy, which had grown convoluted. The United States had supported a United Nations vote affirming the principles of the Geneva Protocol, but it argued that riot-control agents were not covered by the treaty. The protocol, in fact, lists no specific agents, but instead prohibits "the use in war of asphyxiating, poisonous or other gases, and of all analogous liquids, materials or devices" as "justly condemned by the general opinion of the civilised world." Hornig advised the president that the US government should emphasize that it held fast to a no-first-use policy for chemical and biological weapons, as it defined them.

On February 14, 1967, Meselson, Edsall, and several other scientists presented the five thousand signatures and the petition to Hornig at the White House, then held a press conference. Shortly afterward, President Johnson requested the State Department to formulate a "public statement setting forth a 'no-first-use' policy for biological and chemical warfare agents."[36] At the White House, Spurgeon Keeny, one of Johnson's national security advisors, formulated the statement, which declared that the United States had no intention of initiating chemical or biological warfare; the statement, however, exempted riot-control agents and herbicides from the category "chemical weapons." The Joint Chiefs of Staff objected to even this statement's being made publicly by the president, for the military restraints it might possibly justify. Secretary of Defense McNamara decided in favor of the Joint Chiefs and their claim for full latitude in time of war.[37]

Congressional Leadership

Robert Kastenmeier, a Democratic member of the House of Representatives from Wisconsin, was an early advocate for White House reaffirmation of the Geneva Protocol ban on first use of chemical and biological weapons. Part of his argument was that the United States was losing its moral credibility, and thus the propaganda battle with the Soviet Union that was being waged in neutral countries.[38] Kastenmeier was ahead of his time, as were other members of the US Congress, who then began seeking restraints on the Chemical Corps.[39]

In the United States, a sense of immediate public danger from Chemical Corps activities increased sharply from 1967 to 1969 and set the stage for congressional inquiry and action. The secret testing of dangerous agents, notably the nerve agent VX, came to light, along with hazardous Army plans to transport tens of thousands of obsolete nerve gas bombs and rockets by train across the country to be dumped into the Atlantic Ocean.

On March 14, 1966, an Air Force jet plane in a test over Dugway Proving Ground accidentally released gallons of the nerve agent VX over private grazing land. Due to what the Army later characterized as "freak" weather conditions, the agent was blown some forty miles from the source to Skull Valley, where it killed or sickened thousands of grazing sheep. The Army quickly denied responsibility, insisting to the press that open-air nerve gas tests were not ongoing at Dugway. Within a week, the office of Utah Senator Frank Moss released Defense Department documentation of a March 14 outdoor nerve gas test at Dugway.[40] It took a year before congressional hearings exposed the details of the accident, which prompted media comparisons to the anthrax bomb tests at Gruinard Island and potential dangers to civilians nearby.

In addition, early in 1968, revelations about the US field tests in the Pacific with lethal biological agents caused a public controversy.[41] In the *Washington Post,* Nobel laureate Joshua Lederberg observed that any participant in this kind of project would by definition be taking irresponsible risks with public safety. US biological weapons facilities at the time were producing and storing hundreds of pounds of lethal agents. Lederberg, a strong supporter of treaty restraints on biological weapons, asked, "Is anyone competent to play with these matches designed to ignite 'controlled pestilence'?"[42]

Early in 1969 New York Democratic Representative Richard McCarthy, galvanized by a televised account of the Skull Valley disaster, began to press the Pentagon for more public accountability about chemical and biological weapons. In his book on this campaign, McCarthy expressed his main objective, to dispense with dangerous and unnecessary secrecy and to promote public discussion: "There are indications that the secrecy used to cloak our activities in these areas was as much designed to prevent the American people from voicing their beliefs on these matters as it was for purposes of military security. A sharp distinction can and must be drawn between the public issues that can and must be discussed. The technical aspects of nuclear weapons have been as tightly held as any weapon yet the issues of their use and development have been the focus of useful public debate since Hiroshima. No less can be asked of chemical and biological weapons."[43]

Many journalists wrote about chemical and biological warfare issues during the 1960s. Of these writings, journalist Seymour Hersh's 1968 book *Chemical and Biological Warfare: America's Secret Arsenal* had the most public impact. Hersh revealed the post–World War II programs and unearthed new information on their intense expansion during the 1960s. Citing large budget increases over the previous decade, Hersh described "spectacular advances" in biological research, including drug-resistant agents, being invented by a large and secret chemical and biological weapons program at six military bases and involving more than seventy universities (some overseas) as well as private corporations and contractors.[44]

Although McCarthy raised concern about biological agents polluting the environment, the dangers of accidents and emissions at biological weapons program sites never figured as dramatically as the Skull Valley VX incident or resonated with environmentalists.[45] The reported deaths of several workers from anthrax, of hundreds of laboratory infections with pathogenic agents, and the possibility of plague escaping from a facility raised only small alarms when former program workers reported them to the press.

Meanwhile the risks to the public from chemical weapons seemed to multiply. Alarmed by the Skull Valley accident, Denver area residents demanded that chemical weapons stored at nearby Rocky Mountain Arsenal be moved. The deep underground disposal of liquid waste from the arsenal had already caused environmental problems, including an unusual series of small earthquakes. In response to local protest and without consulting the Department of Interior or the State Department, the Army arranged to ship 27,000 tons of chemical agents and weapons by train to Elizabeth, New Jersey, to be loaded on old Liberty ships. Under Operation CHASE (for "cut holes and sink 'em"), the ships would then be scuttled east of Long Island.

Alerted to this plan, McCarthy called for congressional review of the risks to local communities along the train route and the effects of chemical dumping on the ocean environment. The inquiry revealed that Congress was behind the curve. The Army had previously completed eleven munitions dumps of this type, three of them involving poison gas.[46]

The chemical weapons transport plan operated at levels of risk tolerable to the Army but intolerable to the public. In congressional hearings, Defense Department officials dismissed chances of train collisions or railroad crossing accidents or sabotage. Shortly after such denials, an ammunition train carrying tear gas and explosives to be shipped to Vietnam blew up in Nevada. Increasingly beleaguered, the Army asked the National Academy of Sciences to examine its plans for chemical dumping. Headed by Harvard chemist George Kistiakowsky, formerly science advisor to President Eisenhower, the panel conducted inspections of the proposed rail route, the stockpiled weapons and agent and of the ships in New Jersey that were to be loaded with them. The panel concluded that most of the chemical weapons and containers could be taken apart on site and the contents detoxified by chemical treatment.[47] Contrary to the Army's claim that rail transport and sea dumping was the safest way to dispose of the chemicals, the panel concluded that transportation and dumping posed a serious, possibly catastrophic risk to civilians and could damage the ocean environment. In earlier years, a munitions-disposal ship had exploded just five minutes after sinking and another had been lost in rough seas. A particular danger exposed by the panel was the possibility that an underwater ship explosion could release a massive cloud of nerve gas that could be blown over densely populated areas of the East Coast.

Nixon's Agenda

On becoming president in 1969, Richard Nixon appointed Henry Kissinger as his national security advisor, creating a unique relationship for executive review of defense and national security policy. Kissinger, a critic of the Kennedy and Johnson administrations, wanted to combine the intellectual excitement of the Kennedy years with the order of the Eisenhower era.[48]

Toward the close of the previous administration, the Soviet invasion of Czechoslovakia, combined with escalation of the Vietnam War, made President Johnson reluctant to confront his military. By the end of his term of office, the Strategic Arms Limitation Treaty (SALT) talks and Senate ratification of the Nuclear Nonproliferation Treaty had been derailed.[49] For Nixon, these derailments were opportunities to have an international impact. Rather than support the Pentagon's policies on biological and chemical weapons (his stated position as a presidential candidate in 1961), he strongly promoted international arms-control restraints. Nixon and Kissinger were hardly doves. They accepted the doctrine of "assured destruction," which by Pentagon definition in 1968 was one-fifth to one-fourth of the Soviet population and one-half of Soviet industry, and greatly increased US land- and sea-based strategic nuclear capabilities.[50] They were also convinced that it was not Asia but Western Europe that was most at threat, from a war between the superpowers.

For Kissinger, the National Security Council became an important power base, where as chair of the Senior Review Group, he determined what issues should reach the president and when. An adept organizer, Kissinger created committees he then controlled. One was the Verification Panel, formed in July 1969, which managed arms-control negotiations; another was the 40 Committee, in charge of authorizing CIA and other covert activities. In the previous administration, the NSC had a budget of $700,000 and forty-six assistants; under Kissinger, by 1971 the NSC budget was $2.2 million, with more than a hundred people on the staff.[51]

In 1969 national and international opposition against biological and chemical weapons coalesced. Congressional hearings were being held in the Senate and the House on chemical and biological weapons programs. Investigative journalists were uncovering dangerous military accidents and mistakes. The United Nations had become the international forum for the many states now loudly protesting US use of chemicals in Vietnam. As an additional pressure, the US government was also headed for high-level, bilateral arms-control negotiations with the USSR.

Representative McCarthy and others began pointing out that the US chemical and biological weapons programs, combined in the Chemical Corps, had never been subject to thorough high-level review. Since the late 1950s, increases in their budgets had been approved by a few uncritical Congressional advocates and then buried within general appropriation bills, leaving the rest of Congress, the State Department, and the president uninformed.[52] In March 1969, McCarthy asked for an open, joint congressional hearing on chemical and biological weapons. When the Pentagon stonewalled, McCarthy then fired off a list of pointed questions on policy about these weapons to Secretary of Defense Melvin Laird, Secretary of State Rogers, Kissinger, and other senior officials. Laird, who had served in Congress with McCarthy, reacted with an April 30 memo to Kissinger urging a review of chemical and biological weapons and suggesting that the necessary studies be initiated immediately.

Kissinger's NSC staff recommended that he agree to Laird's request, which he did in a memorandum to Laird dated May 9, stating, "I fully agree with your concern about US policy and programs relating to chemical and biological warfare. A study will be initiated shortly in order to facilitate early consideration by the National Security Council."

In late May, Kissinger, acting on behalf of the president, issued a memorandum to the secretary of state, the secretary of defense, and other key officials directing them to undertake a study of chemical and biological weapons policy, programs, and operational concepts, to be forwarded to the National Security Council by early September. The study was to include an examination of "US policy and programs, the main issues confronting that policy, and the range of possible alternatives thereto."[53]

In June, Henry Kissinger asked Matthew Meselson, a former colleague at Harvard, to prepare a position paper on chemical and biological weapons. Meselson and Harvard chemist Paul Doty, who had been a member of President Kennedy's Science Advisory Committee, organized a private conference to discuss the major policy issues that were under consideration in Washington. Attended by a number of senior scientists, it also included an NSC staff member involved in the review.[54] Meselson's position paper, dated September 1969, urged US ratification of the Geneva Protocol.[55] It outlined the danger of strategic biological weapons to civilians and argued that the United States had no need of such arms. "Our major interest," he concluded, "is to keep other nations from acquiring them."

The president had directed that the study of chemical and biological weapons programs and policies be performed by the Political-Military Group within the NSC.[56] The group coordinated studies and evaluations from the relevant departments and agencies of the government and summarized them in a forty-eight-page report dated November 10. After a factual summary of the existing programs and estimates of activities abroad, it presented a list of pros and cons on the numerous military, political, and legal questions at issue.[57]

Only a few of the departmental and agency studies and position papers submitted to the group are available. The President's Science Advisory Committee (PSAC) recommended that the US government renounce offensive biological warfare, destroy its existing biological weapons stockpiles, and pledge to maintain none in the future. It also recommended the production of a new generation of binary nerve agent chemical weapons, arguing that they would be safer than existing stocks in storage and shipment. Another thorough study conducted by the Office of Systems Analysis (OSA), a division within the Office of the Secretary of Defense, cast doubt on the worth to the United States of both chemical and biological weapons.[58] Its authors saw little value in chemical weapons that were in the incapacitating category, since they had limited and uncertain effects and would invite escalation by an adversary. They noted that the surprise value of chemical weapons would diminish greatly after a first use, since masks and protective clothing could provide effective protection. Wearing such equipment, however, would slow the tempo of operations so that the US government might wish to maintain a lethal chemical capability in order to impose a similar burden on an adversary who initiated chemical warfare. The OSA evaluators dismissed biological weapons, if they could be developed at all, as redundant with nuclear weapons and less controllable.

On the issue of "nonlethal" chemical weapons, the Department of State held that the most persuasive interpretation of the Geneva Protocol was that it prohibited all use of chemicals in war except for those widely used domestically for riot-control purposes, and only if these were used for humanitarian purposes, not in conjunction with other weapons in order to kill or wound an enemy. The Department of Defense disagreed, maintaining that the Protocol did not cover the kinds of chemical agents being used in Vietnam at the time.

On the central question of whether to renounce biological weapons, the JCS remained reluctant to give up the option. Secretary Laird was for keeping the chemical warfare program but, as he had already indicated in an interagency memo and a *Washington Post* interview, he might consent to reducing the biological program to defensive research.[59]

On November 18, 1969, the National Security Council met with President Nixon to discuss chemical and biological weapons. JCS chairman Gen. Earl Wheeler was alone in arguing for retention of biological weapons, agreeing only to a no-first-use policy. Before this meeting, the president had already made known that he favored the idea of the biological weapons treaty being discussed at the Eighteen Nation Disarmament Committee in Geneva. The UK draft of the treaty had already been approved by the United Nations General Assembly in 1968 and was favored by NATO (the North Atlantic Treaty Organization). In July 1969 Nixon had communicated personally to Geneva that he approved the draft treaty's principles, along with other initiatives for strategic arms control based on "the wisdom, the advice, and the informed concern of many nations." He stated that the "specter of chemical and biological warfare arouses horror and revulsion throughout the world."[60]

The Nixon Announcement

On November 25, speaking from the White House, President Nixon renounced biological weapons for the United States and put restrictions on further production of chemical weapons.[61] "This is the first thorough review ever undertaken of this subject at the Presidential level," Nixon stated; later in his address he again referred to the review as "unprecedented." Not even an interdepartmental review had been conducted since 1954, although spending and program scope increased greatly, especially after 1961.[62] Henceforth, as he directed in National Security Decision Memorandum 35 dated that day, the US government would renounce the use of "lethal methods" and "all other methods" of biological warfare and would confine its biological programs to "research and development for defensive purposes."

Regarding chemical weapons, the memorandum "reaffirmed" US renunciation of the first use of lethal chemicals and extended the renunciation to incapacitating chemical weapons. The retaliation-only policy for chemicals allowed the Chemical Corps to continue its development of binary weapons, which kept the chemical precursors separate until just before use, thereby reducing risks in transport and storage.

In deference to their continuing use in Vietnam, an exception was made for riot-control agents and herbicides, for which a separate decision memorandum was promised. Late in 1970, Nixon halted the use of herbicides in Vietnam, but the use of riot-control agent was allowed to continue. Although President Nixon sent the 1925 Geneva Protocol to the Senate for its advice and consent to ratification, the Pentagon's continuing authority to use riot-control chemicals in Vietnam would prove a stumbling block for another five years.

In his address renouncing biological weapons, Nixon widened the scope of his remarks, saying, "Mankind already carries in its own hands too many seeds of its own destruction. By the examples we set today, we hope to contribute to an atmosphere of peace and understanding between all nations." Within a month the US and the USSR began the SALT talks, eventually resulting in an agreement to limit deployed strategic nuclear weapons and in the 1972 Antiballistic Missile Treaty.

Toxins

In his 1969 renunciation, Nixon failed to refer specifically to toxins that can be produced by bacteria or other organisms—which, because those toxins are substances rather than organisms, are also chemical poisons. The PSAC committee in its report had defined toxins as chemicals, which the WHO had also done, although not for treaty purposes.[63] The US Army, suffering from the shutdown of its biological warfare programs, was eager to maintain the chemical definition, rather than lose its toxin projects and the resources that went with them, such as the Pine Bluff facility.

A March 1970 memorandum from Matthew Meselson to Henry Kissinger emphasized that toxins had numerous shortcomings compared to chemical weapons that the US government had already developed and stockpiled. Whether toxins were made by living organisms or by chemical synthesis, keeping them as a military option would soften the hard line against biological weapons that the President's decision intended to communicate.[64] Why should the president risk sacrificing a large, symbolically important international gesture for a minor military project. On February 20, 1970, Nixon included toxins in the US renunciation of biological weapons, regardless of their means of production.

Toward a Biological Weapons Ban

Nixon's renunciation of biological weapons also carried with it two decisions regarding treaties. One was that the president would, with advice and consent from the Senate, ratify the Geneva Protocol. The other was that the United States would support the United Kingdom's draft of a treaty to ban the development, production, possession, and stockpiling of all biological weapons; this draft was proposed in the summer of 1968 to the Eighteen Nation Disarmament Committee. The British position upheld the US distinction between chemical weapons, which had already been used in war, and biological weapons, which because they were less developed, might be more easily banned. The UK working paper first submitted to the ENDC called for a new "Convention for the Prohibition of Microbiological Methods of Warfare, which would supplement but not supersede the 1925 Geneva Protocol."[65]

Although the US and UK governments separated the problem of biological and chemical weapons, other nations, notably the Soviet Union, wanted them kept together. US use of riot gases and herbicides in Vietnam provoked international protests against the US failure to ratify the Geneva Protocol, which applied to both chemical and biological weapons use. The USSR was reluctant to let the United States separate chemical from biological weapons in a new treaty.

In 1971, Soviet opposition to the separation of biological and chemical weapons caused the first meetings on the UK treaty to end in a stalemate. The 1972 sessions threatened to prove just as frustrating. Then, in a sudden about-face, the Soviets agreed to separate biological weapons from chemical ones and gave their support to the biological weapons ban. The USSR became, with the United States and the United Kingdom, one of the three depository nations for the treaty.[66] Whether the Soviets intended at that time to create a secret offensive program, which they authorized in 1973, may never be known. The treaty, lacking compliance measures, such as onsite inspections or other verification provisions, allowed the Soviets to make of it what they would or to ignore it. At face value, the USSR in 1972 showed its willingness to participate in future arms-control agreements. As a diplomatic concession on the part of the United States, Article IX of the treaty enjoined each state party to a commitment "in good faith" to continue negotiations to reach a ban to prohibit the development, production, and stockpiling of chemical weapons, a goal achieved in 1993, after the collapse of the Soviet Union.

On April 10, 1972, on behalf of the United States, Nixon signed the "Convention on the Prohibition of the Development, Production and Stockpiling of Bacteriological (Biological) and Toxin Weapons and on their Destruction."[67] Nixon's renunciation of biological weapons halted the expanding US military experiments, the production of agents, and the field tests that undoubtedly would have subsequently exploited the early discoveries in genetics and molecular biology, as well as the later spectacular breakthroughs in biotechnology. The Biological Weapons Convention (BWC), as it is usually called, followed through on Nixon's decision, severing the link between Western microbiology and state weapons programs. By the time the BWC came into force in 1975, Nixon had resigned from the presidency under threat of impeachment because of the Watergate scandal.

Ratifying the 1925 Geneva Protocol

The ratification of the Geneva Protocol, which had stalled in the Senate in 1925, stalled again in the early 1970s, as the US military, still waging war in Vietnam, wanted to keep the options of riot-control agents and herbicides.

The days of the US Army's use of herbicides were numbered, however. In April 1970, the Department of Defense banned the use of Agent Orange (one of three principal herbicides used in Vietnam), responding to concern that exposure to it might cause birth defects. The same year, the American Association for the Advancement of Science approved and financed a pilot onsite investigation of the effects of herbicides in Vietnam. Matthew Meselson led the team, which included physician John Constable and forestry expert Arthur Westing. After receiving cooperation from the American military in Vietnam, they returned with aerial photographs and other documentation showing that herbicides were highly destructive to mangrove forests and that crop destruction, supposedly confined to military food production, was targeting civilian crops. The team interviewed mothers in Vietnamese villages but found no clear evidence for an association of herbicides with birth defects. The day the team reported its findings to the annual meeting of the AAAS in Chicago in December 1970, the White House announced it was ordering an end to the use of herbicides in Vietnam.[68]

In March 1971, six days of hearings before the Senate Committee on Foreign Relations laid out the issues regarding US ratification of the Geneva Protocol.[69] The State Department supported the military's desire to preserve the option to use riot-control agents in war. The opposing argument was that the use of riot-control agents had escalated, showing the need for a total ban in war as a "firebreak" against such proliferation in types of munitions and their use. McGeorge Bundy, then head of the Ford Foundation, testified that the current military utility of chemical agents meant little when balanced against long-term consequences: "[When] we weigh the claims of the future, the scales fly up. In that future there will be other herbicides—other riot control agents—other gases—in other hands. Far more important than any local tactical advantage of tear gas or herbicides is the need for the strongest possible international barrier against the gas war of the future."[70]

In 1975, President Gerald Ford and the US Senate Committee on Foreign Relations reached a compromise limiting the future use of chemicals in war, whereupon the Senate approved the 1925 Geneva Protocol and the president ratified it, half a century after the treaty had been opened for signature.

The Aftereffects

In 1969, Nixon envisioned the defensive biological weapons program's becoming a world leader in immunization against global epidemics, with Fort Detrick best known for the US Army Medical Research Institute of Infectious Diseases (USAMRIID), not preparations for biological warfare. This great unilateral abandonment of an entire class of weapons was as unprecedented as the creation and expansion of this innovative and now failed program. Dismantling the biological program was a massive and often difficult undertaking. The reorientation of Fort Detrick was at first delayed, encumbered by bureaucratic red tape and a lack of coordination between Congress, the White House, and federal agencies. To facilitate the transition, other agencies crucial to change—for example, the Public Health Service—needed White House directions, which were slow in coming.

In 1969 Nixon also signed the National Environmental Policy Act, another of his legacies, which required biological weapons program officers to show in writing that they had met environmental safety standards. Hundreds of pounds of agents, including dried anthrax and tularemia bacteria and nearly 160,000 pounds of wheat rust and 2,000 pounds of rice blast, had to be safely destroyed, along with tens of thousands of filled munitions. Dry anthrax was sterilized and removed from munitions and denatured. Biological weapons facilities had to be decontaminated and equipment dismantled, all according to law. As one report noted, "Extensive press coverage was provided through a constant series of briefings, news releases, closed circuit TV, and tours of non-contaminated areas throughout the operation."[71]

Other changes were necessary. Commercial and academic contracts were cancelled. All crop disease research was transferred to the Department of Agriculture. Research on detection devices and protective clothing was shifted to Edgewood Arsenal and the Chemical Corps. Defensive testing at Dugway Proving Grounds was sharply reduced. With the budget for biological research and development cut in half, some program officials hastily began rewriting their research proposals as defensive rather than offensive in intent.[72]

Even before the BWC came into force, a new phase of the Cold War had begun. With it began a new cycle of secrecy and mistrust that encouraged biological weapons proliferation in the Soviet Union, where few expected it. As the United States abandoned its biological weapons, the Soviet Union covertly violated the convention. Its motivation may have been national defense, but what it stood to gain from secret biological weapons remains unclear. It certainly suspected that the American offensive program was continuing in some form. Following the Nixon decision, the CIA retained small amounts of anthrax and shellfish toxin as well as cultures for tularemia, brucellosis, Venezuelan equine encephalitis, and smallpox. This defiance of the president's authority was made public in Senate hearings in 1975 and likely attracted Soviet attention.[73] US intelligence might have deliberately fomented USSR suspicions of a secret US program through a double agent in the Soviet military.[74] The Soviet war in Afghanistan, its Vietnam, likely increased military latitude in a government where it already wielded significant power.

Whatever its motivation, the Soviet Union purposefully violated the Biological Weapons Convention, and, once revealed, its example of secret deviance undermined credibility in the treaty. After Western powers acquired their nuclear weapons and renounced biological warfare, the Soviet Union soon embarked on a secret project that, like those that went before, was aimed at civilian targets.

CHAPTER 7

The Soviet Biological Weapons Program

The 1972 Biological Weapons Convention completed the codified restraints on germ weapons: not only their use but also any preparations or possession for such use were outlawed. Biological weapons, the only weapon of mass destruction then internationally banned, became legally distinct from chemical weapons, which had a demonstrated if limited tactical potential and some potential for deterrence, as during World War II when neither the Allies nor Germany dared initiate first use. Biological weapons also became legally set apart from nuclear weapons, which had known strategic power and the potential for deterrence against use during the Cold War. After 1975, the dominant issue for those concerned about biological weapons was how and if legal restraints actually prevented secret proliferation. Did any nation have the right to secrecy in the face of strong international consensus against germ weapons?

The Problem of Promoting Compliance

After World War II, the severe military restrictions on Germany and the other Axis powers included bans on biological, chemical, and nuclear arms. With the 1972 Biological Weapons Convention (BWC), sovereign nations agreed to the ban. Yet, formulated without measures to promote transparency, the treaty could not prevent or reveal secret offensive programs if a state party was willing to risk violation. The BWC lacked strong compliance provisions, such as mandatory declarations of past and current capabilities that might be verified, a stable independent organization to oversee and encourage state compliance, and joint exercises or onsite inspections to foster communication among states and build confidence that the prohibition was effective. In the event of an unusual disease outbreak, no standard means were in place for credible expert inquiry. Allegations or complaints might be lodged with the United Nations Security Council, but there Cold War politics were likely to diminish the chance of disinterested inquiry, as they did in 1952 when North Korea accused the US of biological weapons use.

From 1969 to 1975, the United States set the maximum standard for compliance. It fostered high-level civilian review of its military program and policies, unilaterally repudiated biological weapons, and publicly dismantled its large offensive facilities. The United Kingdom also openly dispensed with its biowarfare program; by 1968 it was holding "open days" for the public at Porton Down to emphasize its public health mission. By 1968, France, a nuclear power, had also repudiated its postwar investment in offensive biological weapons and gave its support to the BWC. The Soviet Union, with the United States and the United Kingdom, had formulated the treaty's provisions and was with them a depository nation. The price of Soviet participation was minimal incursions on state and military secrecy. After signing the treaty, the Soviet Union fully exploited this weakness by covertly developing an offensive program that exceeded the capacity the United States had earlier achieved.

As the Soviet Union was collapsing, the secrecy surrounding its biological program broke down. Former program scientists began revealing details of a secret, bureaucratically enormous program, which employed thousands of scientists and technicians and supported large research and production facilities. Like the British and US programs, the Soviet program emphasized experiments on infectious agents and toxins and aimed for large-scale industrial production for strategic attack. Its technical achievements were uneven but intimidating. Without great success, it emulated Western science to increase the power of some germ agents. With time, it sought to add long-range intercontinental missiles to aerial delivery.

Beginning in 1991, exchange site visits involving Soviet, British, and American experts shed more light on the scientific and industrial dimensions of the USSR program. Later, other cooperative ventures added detail to the picture of the large state bureaucracy in which the Soviet program originated and grew, known to relatively few people even within the USSR. It took regime change to bring transparency to the Soviet system, so that the program was basically defunct by the time it was revealed. Nonetheless, the 1990s were marked by increased international experience in onsite inspections and documentation that pointed the way to improved verification and compliance. The most important question the Soviet program raised was how international law could diminish the biological weapons threat posed by a secret state.

The objective of the 1972 Biological Weapons Convention was and is to eliminate the offensive state-sponsored programs and their dangerous products. A major problem in assessing state compliance is the secrecy claimed in the name of political sovereignty. Within that is the specific biological warfare problem of "dual use," the possibility that military activities can be disguised as commercial research, development, and volume production. In the early debates about how to ban chemical weapons, this problem was also central. The commercial production of chemical compounds for industry required laboratories and plants that could be converted to military use, for example, German dye works in World War I. Some of the most notorious chemicals (such as chlorine and phosgene) also had peaceable industrial purposes.[1] With biological weapons, dual use often refers to large laboratories and pharmaceutical plants, where either legal or prohibited agents might be prepared and grown in bulk. In contrast to chemicals, the agents historically developed for biological weapons have had little or no commercial value that would make it necessary to produce and distribute them in great quantity. Many are also easily found in nature. Even the detection of traces of a dangerous agent such as anthrax or tularemia at a site can raise suspicions, which might be founded or unfounded.

The microbiologist's skills have also been included as part of the problem of dual use. The laboratory scientists who developed biological weapons were generally trained for academic, public health, or commercial employment. What distinguished them was their employment in state programs.

A state-sponsored program may not be, after all, easy to hide. It requires a considerable physical plant for handling and storing dangerous pathogens; verifying their virulence in animals; culturing and concentrating them in quantity; safely disposing of wastes; and developing and testing munitions or other devices in chambers. It also requires outdoor space for field tests of munitions and simulations of attacks. Further, a biological warfare program generates a paper trail. It relies on trade transactions, perhaps international ones; official records; research documentation; the recruitment and training of personnel; organizational integration into military units; and a variety of other activities. If successfully halted, any one or several of these functions could stop a program cold.[2]

The secrecy and disinformation that obtained within the totalitarian Soviet Union, though, combined with its resistance to outside contacts and its enormous size, successfully obscured signs of its large program. Cold War intelligence on Soviet biological weapons was necessarily faulty and uncertain. As the USSR program grew, relatively few people even there knew how the risks of proliferation were multiplying.

The Origins of the Soviet Program

The historical overview of the Soviet biological weapons program has generally been pieced together from multiple sources, primarily those from German sources documented following World War II, from former Soviet biological weapons scientists as the USSR collapsed, from scholars working with newly available archives, and also from recent East-West scientific exchange and demilitarization programs.[3] This information allows a composite sketch of both the old biological weapons program from early in the century and of the later Soviet offensive initiative, called Biopreparat, that began soon after the 1972 Biological Weapons Convention was signed in Washington, London, and Moscow.[4]

In 1945, Allied forces captured two Germans who had access to intelligence information about the Soviet program. Col. Walter Hirsch, an Austrian chemist in the German chemical weapons program, wrote a long report for US military intelligence on the USSR chemical program that also includes information on the biological weapons program.[5] Another German official, Dr. Heinrich Kliewe, from the German biological program, was also debriefed by US Army intelligence on what he knew about the USSR program.[6]

In late 1989, Soviet program scientist Vladimir Pasechnik defected to the United Kingdom. Although he left no comprehensive overview, information he relayed about Biopreparat, especially research at his Leningrad institute, was the first intimation of the scope and offensive intent of the Soviet venture. In 1992, the former deputy director of Biopreparat, physician Ken Alibek, defected from Russia to the United States; his debriefing and, later, his 1999 memoir of his military career added much contemporary detail. Also in the aftermath of the Cold War, retired microbiologist Igor Domaradskij provided information on the program from the perspective of a high-ranking civilian scientist.[7]

In the period following World War I, the Soviet Union carried out and published extensive research on bacteriology in the institutes of the People's Health Commissariat created to stave off recurrent epidemics that were ravaging the country. Russian scientists were deeply influenced by French and German microbiology of the previous century and, relying on research institutes throughout the nation, made important contributions of their own to the control and prevention of infectious diseases such as anthrax, plague, and cholera.

Like other industrialized military powers, the Soviet Union organized its initial biological weapons program on the basis of its chemical program. After World War I, when both Germany and the USSR were politically isolated from Western Europe, they shared air force training and the testing of chemical weapons, ostensibly for defensive purposes.[8] One of the participants in these joint exercises was Jacov Fishman, in 1925 the head of the Soviet Military Chemical Agency (MCA). In 1928, Fishman, who is credited with being the father of Soviet biological weapons, proposed the basic framework that gave the MCA offensive responsibilities and delegated defensive efforts to the Institute of Chemical Defense and the Ministry of Health.[9] Fishman's innovations were part of the modernization of the Soviet military which, under the leadership of Marshal Mikhail Tukhachevsky, fostered central command and technical innovation based on Western models.[10]

The joint Soviet-German chemical and air force exercises ceased after Hitler came to power in 1933. The much-publicized 1934 Wickham Steed report of German experiments with biological agent simulants in Paris alarmed the Soviets as well as the British and French.[11] The Soviets also feared that the technologically advanced British might resort to bacteriological weapons, which its officials derided as "a novel type of warfare prepared in bourgeois countries."[12]

Col. Hirsch's principal informant was a captured Soviet scientist who, starting in 1933, had worked at the Scientific Medical Institute of the Red Army in Moscow. By this prisoner's account and others, the Soviet biological weapons program in the late 1930s moved aggressively to establish projects in multiple research institutes near Moscow as well as in the Urals, where during the war military industries were being built beyond the reach of German troops. For example, in Chkalov (Orenburg), a bacteriological research institute appears to have been concerned with the dissemination of bacteria in air and in 1943 another bacteriological facility was built in Sverdlovsk (Ekaterinburg), for military purposes.

Fishman early on proposed that open-air field trials for biological weapons be conducted in remote locations within the USSR. In 1936, Ivan Velikonov, director of the Scientific Medical Institute, which was under Fishman's jurisdiction, led the first of such trials at Vozrozhdeniye (Rebirth) Island in the Aral Sea. Complete with two ships and two aircraft, the trials apparently lasted from spring 1936 to fall 1937.[13]

In 1937, both Fishman and Velikonov were arrested and exiled in Stalin's purge of the military, which extended to public health and medical officials.

Many biologists, including the heads of institutes, were arrested on charges of sabotaging the health of the state, and in show trials they confessed to committing sabotage for Japan or Germany by spreading hog cholera, causing anthrax deaths among cavalry horses, and poisoning military food supplies. Also in 1937, Marshall Tukhachevsky, an avid proponent of combined chemical and biological capabilities, was executed for treason.

Soviet biological warfare activities after the 1937 purges are less well documented than the declining state of the state's biological science research. During the war, Stalin began supporting Trofim Lysenko, the agronomist who theorized that the environment could change hereditary traits and denounced modern genetics. Lysenko's influence, like that of racialist scientists in Nazi Germany, spelled a significant setback for scientific research and it lasted for decades.[14]

The Soviet's foremost military advocate of biological weapons was Gen. Yefim Smirnov, head of military medical services during World War II and, after the war, the Soviet Minister of Health. Known as "an impassioned advocate of biological weapons," in 1973 he took charge of the Fifteenth Directorate of the Ministry of Defense, the organization that was intended as Biopreparat's "customer" for these innovative weapons. Smirnov retired from this position in 1981 but remained a figure to be reckoned with until his death in 1989.[15]

Yuriy Ovchinnikov, an accomplished biochemist of the next generation and later vice president of the Soviet Academy of Sciences, is often described as the scientific visionary behind Biopreparat. In the early 1970s, he reportedly convinced Soviet Secretary General Leonid Brezhnev and other authorities that the USSR seriously lagged behind Western genetics and molecular biology and should begin a national research program to close the gap. Ovchinnikov may have intended only scientific advances that had pharmaceutical value, without any exploitation by the military. The totalitarian government of the USSR made this impossible. In 1973, the Soviet Union created Biopreparat as a conglomerate for commercial research and a mask for its secret biological weapons program. Under the Council of Ministers, the USSR's highest governing body, Biopreparat involved the Ministries of Health, Agriculture, and Chemical Industry, as well as the USSR Academy of Sciences and the Academy of Medical Sciences and the KGB. This biological weapons program relied on a dozen research and production centers throughout the country. It employed some nine thousand scientists and technicians, although only the upper echelons fully understood the offensive purpose of its mission. The size of a small city, Obolensk, a closed biological weapons compound outside Moscow, was built in the 1980s along models developed years before for nuclear-weapons development and production. Post–Cold War revelations about the Soviet program's capabilities, for instance, the genetic manipulation of pathogens and the capacity to produce hundreds of kilograms of anthrax spores monthly, created great concern in the early 1990s.

Biopreparat's top officials and many of its employees were drawn from the Army's Fifteenth Directorate, which exercised great authority in influencing and expanding the Soviet biomedical infrastructure. Biopreparat's Interagency Scientific and Technical Council, which assembled representatives from other bureaucracies, including the KGB, was its "brain center," where decisions were made on projects.

As in other biological weapons programs, Biopreparat's main functions were laboratory research to guarantee pathogen virulence and large-scale production for munitions. Using older Western techniques, Soviet scientists experimented with bacteria resistant to antibiotics, antigen-altered bacteria and viruses, and enhanced survival of microorganisms in the environment, for example, improved resistance to sunlight and moisture.[16] Soviet scientists also developed antibiotic-resistant strains of plague and tularemia and attempted to combine pathogens genetically to increase their virulence.

The system supported scientific institutes in Moscow and Leningrad. Other facilities were located at Kirov, five hundred miles east of Moscow, and in Siberia, Uzbekistan, and Estonia. Biopreparat created new biological weapons enclaves, such as the one at Obolensk and another at Koltsovo in Siberia (known now as Vektor).

Biopreparat built factories for biological-agent production, most impressively a large plant at Stepnogorsk in Kazakhstan, and six other facilities that reportedly could quickly produce large amounts of agents if mobilized.[17] For open air testing, the USSR military relied on Vozrozhdeniye Island and other remote sites, just as Jacov Fishman had advised, and maintained four facilities for development and production, including the one at Sverdlovsk.[18] The extent to which biological weapons were assimilated by the Soviet military remains an open question. According to Ken Alibek, if word came from Moscow, the entire system stood ready to produce pathogens, load them into munitions, and launch them abroad. [19] But whether high-level Soviet military commanders actually had war plans incorporating biological weapons, for example, for surreptitious use against Afghanistan or in a last-ditch attack in a nuclear war with the West, is unknown.

The motivation for the Soviet violation of the BWC may never be fully understood. Leonid Brezhnev and civil and military leaders who knew about the program might have perceived a military advantage in developing a type of weapon in which other powers had no interest, perhaps to be used in combination with nuclear, chemical, or other weapons. Or the Soviet initiation and expansion of the program may have been a bureaucratic response to the Cold War arms race, a flaunting of the 1972 BWC but completely in keeping with the destructive power of nuclear weapons. By some insider accounts, the Soviets never believed that the United States terminated its offensive program in 1969. Therefore, it sought to keep pace with an American weapons program it suspected was drawing on the major science breakthroughs reported in Western journals.[20] During the 1980s, US defensive research projects were known to involve genetics and cloning and to include a long list of exotic pathogens (including Ebola virus, Marburg virus, and Rocky Mountain spotted fever) and exotic toxins that posed minimal public health threats but could be lethal biological weapons.

After 1975, American suspicions about a Soviet offensive program waxed and waned. In 1978, when a Bulgarian exile, Georgy Markov, was assassinated in London by a passerby using the toxin ricin, the US government suspected that Soviet intelligence had technically assisted in this assassination. The suspicion was correct.[21] But was the Soviet Union capable of concerted use? And how could that be determined?

The "Yellow Rain" Allegation

President Jimmy Carter entered office in 1976 with hopes for bilateral arms-control negotiations. Those hopes disappeared in December 1979, when the Soviet army invaded Afghanistan to reestablish a pro-Soviet government and found itself mired in a long and unpopular war. During Carter's administration, rumors began circulating from Southeast Asia about a mysterious and deadly "yellow rain" affecting mountain Hmong tribes, allies of the United States during the Vietnam War who had run afoul of the Laotian government. In 1981, after Ronald Reagan became president, the administration began capitalizing on the yellow rain rumors to bolster its confrontational approach to Cold War politics. High on its agenda was support for the army's renewed production of chemical weapons.

In September 1981, in a speech in West Berlin, Reagan's Secretary of State Alexander Haig accused the Soviet Union of abetting the use of lethal trichothecene mycotoxins in Southeast Asia.[22] This accusation was the beginning of a Cold War controversy that resembled the 1952 Korean War allegations. But now it was the United States that was making the accusation, piecing together its case from intelligence and other sources. Also unlike the Korean case, the circumstances of this controversy allowed significantly higher standards for independent expert inquiry.

Laos, Vietnam, and Cambodia were inaccessible, and Afghanistan, which figured in a minor way in the accusations, was in the midst of a war with the Soviet Union. Access to alleged witnesses was nonetheless possible. Hmong refugees had been fleeing Laos for years, and many were living in refugee camps in Thailand. These refugees were in fact the primary source of accounts that Vietnamese or Laotian planes had sprayed their villages and fleeing refugees with a lethal substance that caused sickness and death. When asked for samples of this material, the Hmong provided vegetation and stones with yellow spots on them. Eventually, a university laboratory under US government contract conducted chemical analyses of these and other samples and reported mycotoxins (toxins produced by fungi) as the yellow rain agent. The physical samples, the refugee interviews, and the US government's chemical analyses were each available for independent expert review.

Standards of Evidence

Consulted by the CIA in 1980, Harvard biologist Matthew Meselson was intrigued by the variety and ambiguity of the yellow rain evidence.[23] In 1983 he organized a conference of academic and government experts in Cambridge to discuss the evidence thus far accumulated—about the nature of the toxin agent, where it was found and reported, what scientists had tested it and what witnesses had described the airplane attacks. The first part of the US accusation to unravel was the discovery that the yellow spots on leaves and stones were, in fact, the innocuous feces of Asian honeybees.

The US government had itself conducted chemical analyses of the physical samples but neglected to examine the material under a microscope and to make comparative analyses. Using an electron microscope, botanist and pollen expert Joan Nowicke at the Smithsonian Institution examined yellow rain samples brought to her by the Army and samples sent to her by Meselson, which he had obtained from the Army, the Canadian Department of Agriculture, and ABC News. Her results showed that these samples were largely composed of pollen identical to that in bee feces collected in Southeast Asia in the same ecological environment as the alleged attacks.[24] Onsite in northern Thailand in March 1984, Meselson and bee experts Thomas Seeley and Pongtep Akratanakul documented the common but often ignored phenomenon of yellow "showers" from thousands of simultaneously defecating bees.[25] Comparison of these field samples with yellow rain samples showed them to be virtually identical.

Scrutiny then turned to the Hmong interviews. This testimony turned out to have been informally conducted in the refugee camps by untrained medical staff and US military personnel. Interviews under the auspices of the United Nations and the Canadian government added to the sum total of accounts, which, when analyzed, showed great inconsistency in descriptions, often second hand, of the alleged airplane and rocket attacks and their medical effects on victims.[26]

In 1983, the US Department of State and the Department of Defense authorized a joint team to crosscheck Hmong interviews in the refugee camps. During its two years in the field, the team was unable to corroborate attack accounts.[27] Documents later declassified showed that the team ultimately ascribed the principal attack scenario to "insects or some other natural phenomena."[28]

Secretary Haig's September 1981 Berlin speech had relied on evidence for trichothecene mycotoxins from a single uncorroborated analysis of a leaf and stem sample. Months before the speech, that sample had come not from Thailand or Laos but from Cambodia, accompanied by a muddled story about soldiers falling ill from polluted drinking water. This sample was sent to the Army Chemical Systems Laboratory, where it was divided, with one part sent to toxicologist Sharon Watson of the US Medical Intelligence and Information Agency, housed at Fort Detrick. Watson then sent her sample to plant pathologist Chester Mirocha at the University of Minnesota.

On August 17, in a secret message to the intelligence community, Watson announced that Mirocha had found trichothecene mycotoxins in the material, a type of toxin not on any biological weapons agent list, but she wanted scientific corroboration. On August 31, Richard Burt, then director of the State Department's Bureau of Politico-Military Affairs, instructed a reluctant Watson to prepare a public statement. Her memo on Burt's phone call noted, "Despite these concerns [about corroboration], it was decided to release the information since we were told that it would break anyway with or without our permission."[29]

Mirocha later reported more mycotoxins on other samples, but few questions were asked about his test methods, particularly about controls and false positives, or about the background presence of naturally occurring mycotoxins in Southeast Asia. Nor was any account made public of the results, positive or negative, of the much larger number of additional alleged attack samples and controls he tested.

Following Haig's speech, Britain, France, Canada, Sweden, and the US Army itself commenced a series of careful tests on yellow rain samples. The Army and other government laboratories relied on a powerful and elaborate technique combining gas chromatography and high-resolution mass spectrometry (GC/MS), at a level far more stringent than that employed either in Minnesota or at Rutgers University in New Jersey, where an ABC News–acquired sample had been reported positive.[30] No other methods at the time were likely to be reliable and, without rigorous precautions, even GC/MS could easily yield false positives.[31] Mirocha's results at Minnesota were not replicated; entirely negative findings from the US Army and from Porton in the United Kingdom that undermined the US case were kept secret for years. Negative results from Porton Down, for example, were withheld for four years.[32] Some US Army data remain secret, which suggests that they may not support the US accusation. Meanwhile, when stringent methods were used, laboratories in Canada, France, and Sweden found no positive evidence for an association of trichothecene mycotoxins with yellow rain or with the blood and urine samples from alleged victims.

Mycotoxins, banned under the BWC, were sufficiently ambiguous to allow the Reagan administration to conflate them with chemicals, so that the yellow rain allegations became background noise for support of renewed US production of chemical weapons. The Army wanted to produce binary artillery projectiles and a binary glide bomb called "Big Eye." These projects were resisted by the US chemical industry, which, after Vietnam, wanted no association of its products with warfare and environmental destruction. Chemical weapons were then being used by Iraq in the Iran-Iraq war, to international criticism; the Reagan administration, with its support of Iraq and the Chemical Corps, was then in no position to protest. Ultimately, the Senate was evenly split in its votes for production of binary projectiles and for the Big Eye bomb. In both instances, Vice President George H. W. Bush—later, as president, an advocate of the 1993 Chemical Weapons Convention—cast the tie-breaking vote in favor. With the Soviet Union signaling that it wanted to end the arms race, Congress declined to fund these chemical projects.

The Sverdlovsk Anthrax Allegations

In all biological weapons programs, disease risks posed a hazard for workers.[33] Civilians were also at risk for accidental exposure. The Soviet Union experienced two such accidents. In 1972, as the USSR was involved in BWC negotiations, an epidemic of smallpox broke out in a town in Kazakhstan, not far from a biological weapons testing site. Thirty years later, a former public health minister disclosed that the military had caused the outbreak, although the details of the event remain unknown.[34]

In March 1980, learning of a large outbreak of anthrax in the Soviet city of Sverdlovsk, the US government publicly voiced its suspicion that the Soviets were violating the BWC. In response, Soviet officials insisted that the epidemic was caused by anthrax-contaminated bone meal in livestock feed and the consumption of meat from infected animals. They pointed out that anthrax epizootics (outbreaks among animals) were common in the Sverdlovsk region, which is in the southern Urals, near Siberia. [35]

Matthew Meselson was again consulted by the US government and, along with other experts, attempted to evaluate the ambiguous intelligence information about the anthrax outbreak. He found the Soviet explanation plausible but unproven and sought to conduct an onsite investigation. In 1992, after the demise of the Soviet Union, he was able to lead a team of American and Russian scientists to the scene of the outbreak in Ekaterinburg, the former Sverdlovsk. There the search began. The team sought epidemiological evidence through interviews with the families of victims, eyewitness accounts by medical personnel, and hospital records. Fortunately, the two pathologists who conducted most of the autopsies in 1979, Dr. Faina Abramova and Dr. Lev Grinberg, had preserved tissue samples and case notes from confiscation by the KGB, and this material proved essential in showing that victims died of inhalational not gastrointestinal anthrax. Even more conclusive, epidemiological maps showed that on April 2, 1979, an aerosol emission from a local military facility, called Compound 19, caused around seventy deaths and at least a dozen illnesses among the approximately five thousand people in the area who were exposed.[36]

The team's speculation was that the aerosol release was accidental, due to faulty air filters. The Russian military claimed it had no records of the incident, but Ken Alibek in his memoirs recalled rumors of the military's careless failure to replace an air filter at the facility.[37] The gathering and analysis of information about the Sverdlovsk outbreak, which took nearly two years, showed that a controversial outbreak could be scientifically traced to its source—provided experts were given access and were relatively free from political pressures.

The 1992 Sverdlovsk inquiry also yielded at least two important medical findings about inhalational anthrax, a rare disease not usually well documented. One finding was that its incubation period for some victims might be longer than previously thought. It was usually predicted that infection by an anthrax aerosol led to symptoms within two to five days, followed by nearly certain death within two days after that. (In the Japanese experiments at Unit 100, Chinese prisoners had become sick and died even more quickly.) In Sverdlovsk, the fact that a small number of people became symptomatic weeks after exposure suggested another model, namely, that the anthrax spore in some cases can remain dormant in the human lung for some time, as long perhaps as six weeks, after which it might germinate and cause a fatal infection. During the Sverdlovsk outbreak, once symptoms began (whether two days or six weeks after exposure), death usually did occur two or three days later. Prompt administration of antibiotics could prevent death; Soviet physicians reckoned that they saved fifteen patients this way in 1979, although missing medical records make full case details impossible to confirm.

The 1992 investigation also found that older people were more vulnerable to the disease than younger ones. Many children and young people were in the contaminated zone, yet no one under the age of twenty-four died or appears to have even contracted inhalational anthrax. The older ages of those who contracted inhalational anthrax during the 2001 anthrax attacks supported this finding. The highly unusual Sverdlovsk epidemic is frequently cited as a model for how a covert terrorist attack on an urban population might unfold, eliciting an emergency public health response, from the screening of possible cases to intensive care to the distribution of antibiotics.[38] Using a live spore vaccine called STI (different from the US cell-free vaccine), public health officials from Moscow also planned to vaccinate fifty thousand people in the neighborhood near the military facility. Many residents grew wary when they witnessed the vaccine's side effects, which can include a large ulcer at the point of vaccination and flulike symptoms lasting several days. Around twenty thousand people failed to present themselves for vaccination.

The worst consequence of secrecy and disinformation in Sverdlovsk was the delay in the diagnosis of cases. City officials, ignorant of the military's anthrax project, were taken unaware by the first sudden, mysterious deaths. A week passed before laboratory tests confirmed the diagnosis of anthrax. By then more than a third of those infected, who might have been effectively treated with antibiotics, had died or were dying. Warnings about infected meat misrepresented the true source of danger, which was the military facility. As many as seventeen residents and workers died before they could be hospitalized.[39]

In May 1992 Russian Federation President Boris Yeltsin stated publicly that the military was responsible for the Sverdlovsk anthrax epidemic and decreed that the families of victims should be compensated, a promise left unfulfilled. Further, the language of his decree made it impossible for the military to be held legally accountable for the anthrax deaths.[40] No civil or military hearing was held and, within the Russian military and government, disinformation and misinformation about the cause of the 1979 Sverdlovsk outbreak persisted.[41]

Confidence-Building Form F

In 1990, following Vladimir Pasechnik's revelations about the Soviet biological weapons program, the US and the UK exerted pressure on President Mikhail Gorbachev to admit the USSR had an offensive program. Gorbachev responded with an invitation for them to pick four sites that they would be free to inspect. Intent on verifying Pasechnik's account, they chose four research institutes he had described (in St. Petersburg, Obolensk, Chekhov, and Koltsovo) rather than the controversial Compound 19 in Sverdlovsk. In exchange, a Soviet delegation visited Fort Detrick, Dugway Proving Ground, the Pine Bluff Arsenal, and the Salk Institute in Clearwater, Pennsylvania.

These and subsequent exchange visits went on until 1995, when communication broke down, mostly due to military resistance on the Russian side and resistance from pharmaceutical companies on the US side. Nonetheless, optimism prevailed among parties to the Biological Weapons Convention that Russia would now follow voluntary treaty compliance measures agreed to and modified at the BWC review conferences held after 1975. These agreed measures eventually required six types of declarations, including of all high containment facilities, of defensive research and development programs, of unusual disease outbreaks, of legislation relevant to the treaty, of past activities in offensive and defensive biological research and development programs, and of human vaccine production facilities.

Each declaration has a standard form, with a letter for each category. Confidence-building Form F related to past activities and it was this declaration that was most anticipated from Russia. Early in 1992 President Boris Yeltsin denied any continuation of the Soviet program. In sum, the Fifteenth Directorate had been abolished, and Biopreparat was to become a commercial pharmaceutical venture specializing in diagnostic technology.[42]

In July 1992 the Yeltsin government submitted Russia's confidence-building Form F to other parties to the convention via the United Nations. This document briefly described Soviet and Russian offensive and defensive efforts from 1946 to March 1992. The declaration recounted that at the end of the 1940s the USSR had a defensive biological weapons program. Later, in the 1950s, facilities at Sverdlvosk and Zagorsk (now Sergiev Posad, near Moscow) were constructed for large-scale agent production. These two facilities, plus one at Kirov (now Vyatka), were described as places where work was done on anthrax, tularemia, brucellosis, typhus, plague, and Q fever. Another facility, the Scientific Research Institute of Military Medicine in Leningrad (now St. Petersburg), was also identified as an important research center.[43]

The declaration stated that stockpiles of biological agents were never "prepared or stored" at these facilities, but it affirmed that the plants had been built for mass production of biological warfare agents. Obviously without "justification for prophylactic, protective or other peaceful purposes," in the words of the BWC, the facilities for mass production of biological warfare agents were a gross violation. By inference, biological agents had been tested for virulence, munitions had been designed and tested, and, again, by inference, a doctrine for use had been created. According to the declaration, no stockpiles were ever created and "accordingly there never has been and is not now any problem of their destruction." At Vozrozhdeniye Island, the document affirmed, tests were conducted on experimental agents installed in mockups of airborne and rocket-propelled missiles and spraying devices. Tests were also conducted on warning devices and technologies for identifying infectious diseases.

Form F described 1986 as the year when the USSR began dismantling its offensive facilities, with Biopreparat transferred from the Council of Ministers to the Ministry of Medical and Microbiological Industries. By April 1992, the testing facility at Vozrozhdeniye Island was closed, the conversion of facilities to peacetime uses began, and the biological weapons program was declared terminated. Concerning toxins, Form F noted that the program had investigated botulinum toxin and then ended with this terse conclusion, certainly in reference to the US yellow rain allegation: "In the opinion of experts, mycotoxins have no military significance."[44]

Controlling Scientists and Pathogens

In 1988 and thereafter, as the United States and the Soviet Union reached agreements on nuclear arms control and the Cold War ended, the problem of Soviet compliance with weapons treaties receded. Accords between the United States and Russia continued to emerge and public anxiety about a nuclear holocaust, which had risen during the Reagan administration, subsided.[45] The threat of major powers ever again using chemical weapons also abated. In 1993 in Paris, with support from President George H. W. Bush, more than a hundred nations signed the Chemical Weapons Convention (CWC). This treaty came with compliance measures that the Biological Weapons Convention lacks: an organization based in The Hague, teams of experts, and provisions for mandatory declarations, inspection, and stockpile destruction. With the Cold War ended, and with Yeltsin's stated cooperation, the former tension around Russian treaty compliance was much reduced.

Dismantling the Soviet biological weapons program remained a complex problem.[46] Questions were raised as to whether Gorbachev or Yeltsin had full authority over powerful Biopreparat directors and their military counterparts. During the 1990s, enterprising plans to convert old Soviet biological weapons facilities to pharmaceutical production sprang up and then foundered. A decade after Yeltsin's declaration that Russia had no biological weapons program, the three major military facilities at Yekaterinburg, Sergiev Posad, and Vyatka remained closed to outside observers. The virology center at Sergiev Posad had been the largest and most secret installation and was reported to have experimented on the smallpox virus.

Ensuring openness so that Russia would not revert to an offensive biological weapons program was another main problem. A third fear was that former Soviet biological warfare scientists might be recruited to programs in lesser states hostile to American interests. In 1991 Senator Sam Nunn and Senator Richard Lugar took legislative initiative with the Nunn-Lugar Cooperative Threat Reduction Act. This bill funded the conversion and destruction of former nuclear, chemical, and biological facilities and joint projects with Russian former weapons scientists. The US National Academy of Sciences helped to select and guide worthwhile collaborative projects in biology; the Centers for Disease Control, the Department of Energy, and the National Institutes of Health joined in the effort.

Another perceived threat from the Soviet program was that dangerous pathogens would proliferate. A strain of anthrax developed by Russian scientists proved in animal tests to be resistant to the standard Russian vaccine, although not to a modified vaccine.[47] Another dangerous pathogen was smallpox. Following the 1980 eradication of smallpox by the World Health Organization (WHO), only the Soviet Union and the United States were permitted to have the smallpox virus; the former kept its strains at an institute in Moscow, the latter at the Centers for Disease Control in Atlanta. In the 1980s, the Soviet Union had secretly moved its smallpox reserve to Vektor in Siberia, ostensibly to guard it against terrorists. Failure to inform WHO of this change at the time rankled some in the West. According to Alibek, whose account was disputed, as late as 1990 Vektor scientists had been developing genetically altered strains and had the capacity to produce vats of the virus.[48] The institute preserved fifteen thousand isolates of pathogenic viruses, including some hundred and twenty strains of smallpox, fifty of them unique to Russia. The possibility that loose security would allow unauthorized persons to steal or buy one of the Russian strains was worrisome, especially since smallpox vaccinations had long ago ceased throughout the world.

Some critics felt that security at Vektor was inadequate and that the facility remained ominously secret.[49] Still others felt that joint projects with the West were the best way to reduce the risks of proliferation.[50] Starting in 1999, Vektor and the CDC together embarked on a series of smallpox research projects endorsed and supervised by the WHO and funded at over $4 million by the United States Department of Health and Human Services (HHS) and the Defense Threat Reduction Agency (DTRA) within the Department of Defense.

With reforms, collaborative projects, and the gradual replacement of Soviet-era figures with younger political actors, the biological warfare security risks associated with Russia were diminishing, although not with ideal consistency. Suspicions of Russian motives and handling of US funds for threat reduction persisted, and there remained much to learn about cooperative programs.[51] Still, an important change had taken place: biological weapons were finally uncoupled from the strategic programs of the two major Cold War adversaries and, indeed, from all major state militaries.

The central problem became protecting civilians from biological weapons proliferation from any source and especially from hostile developing nations and terrorists. A reinforced Biological Weapons Convention was first on the list of potential restraints. In 1991 Western European nations took the lead in proposing mandatory compliance measures, analogous to those being proposed for the new Chemical Weapons Convention.[52] The United States, protective of its military and industrial secrecy, balked at required inspections and other measures promoting transparency. In 1994 an Ad Hoc Group was created to formulate a legally binding protocol for the BWC that would overcome these objections by the United States and other countries. The process would prove difficult.[53] The question was whether the United States could balance its new role as the lone superpower in international affairs with its national defense policies.

CHAPTER 8

Bioterrorism and the Threat of Proliferation

As the Cold War was ending, new hopes emerged for a strengthened Biological Weapons Convention through a protocol with strong compliance measures. The 1993 Chemical Weapons Convention offered a model of team inspections, mandatory verification procedures, and a standing organization, located in The Hague. Why not similar fortifications for the BWC?

From the beginning, the United States proved wary of multilateral accord that would require transparency and subject it to international law. In 1991, it requested that a range of verification measures should be reviewed before protocol negotiations began. Two years later the Ad Hoc Group of Governmental Experts produced a technical and scientific report evaluating offsite measures (such as declaration of facilities, surveillance of publications, legislation, and trade, remote sensing and tests of physical materials) and onsite measures (exchange visits, inspection of buildings and key equipment, collection of relevant medical data, and continuous observation of certain facilities, whether by instruments or by experts).[1] In this 1993 Verex (Verification Experts) Report, onsite measures were seen, especially by the United States, as a threat to the commercial proprietary information (CPI) of pharmaceutical and biotechnology companies and to the US defensive biological program. Russia and China also expressed resistance.

Still, since compliance was the goal, not radical transparency, it was agreed that common ground among participating nations might be found. In 1994, an Ad Hoc Group was established to start the negotiations for a BWC protocol that would strengthen compliance measures.

President Bill Clinton supported a protocol for the treaty. But in 1995, after Republicans took control of Congress, the United States became increasingly antagonistic to international treaties, including the 1993 Chemical Weapons Convention, which President George H. W. Bush had promoted and signed. Senate leaders, especially the powerful Jesse Helms, continually objected to the compliance measures being proposed in the Ad Hoc Group negotiations. Although Russia and China also maintained their hesitancy, in the seven years of negotiations their delegates moved toward concessions while the United States introduced a series of proposals that would restrict declarations and inspections in favor of secrecy.

During this same time, biological weapons began taking on new significance for their potential use by terrorists in attacks on American cities. As this chapter describes, there was enough evidence from the Cold War years to indicate that small states, whether allied or neutral, could develop secret biological weapons and should be integrated into the larger world community that sanctioned biowarfare. National defenses against terrorism in general and bioterrorism in the specific were not necessarily in conflict with the BWC protocol. But bioterrorism soon enough became a diversion from a multilateral agreement to curb state programs. Analysts began describing how America's enemies might acquire nuclear, biological, and chemical weapons (NBC), the unconventional weapons developed by the major powers, through the global marketplace and pose an asymmetrical national security threat to unprotected civilians.[2]

This reconfiguration of national defense was part of the US struggle to position itself as the lone superpower with vast military resources in the globalized post–Cold War world, with no Soviet adversary. Regional conflicts in the Balkans, Africa, and the Middle East were generating unpredictable military and political risks, including genocide in the former Yugoslavia and in Rwanda. Nuclear deterrence had a reduced significance in this complex world, as did a US army positioned for Cold War hostilities.

No bioterrorist assault on an American city occurred in this unsettled period. Terrorist attacks, though, were real and demanded new policies. The scale of bomb attacks at home and abroad drew no distinctions between political figures and innocent bystanders, including children.[3] In conjunction with fears of terrorism, anxieties about the risks of bioterrorism increased and, in imagination, mass killings were magnified by epidemic scenarios. Few people had worried about biological weapons in years, yet gradually their potential, refitted to terrorists, generated federal policies that presumed that a state of emergency preparedness should be normal.[4] As the millennium approached, influential politicians and consulting experts broadcast apocalyptic visions of thousands, even hundreds of thousands of Americans dying from unnatural, intentional epidemics of anthrax, smallpox, or some newly devised disease, invisibly inflicted by barbarous foreigners.

Civil Defense Against WMD

Government emphasis on the vulnerability of civilians to weapons of mass destruction (WMD) raises the difficult question of how and if officials expect to protect the population and whether they intend to do this by technology or political initiatives. In a 1961 televised address, President Kennedy warned that the United States was on the brink of a nuclear war with the Soviet Union. To reassure the public, he promised that every American household would receive a pamphlet on how individuals could protect themselves by stocking food and water and seeking fallout shelters.[5] By the time the pamphlet was mailed a year later, the White House and Soviet leaders had confronted and resolved the Cuban missile crisis, to the public's relief.[6] Soon after, in 1963, the United States, the United Kingdom, and the Soviet Union signed the Limited Test Ban Treaty. This accord failed to halt proliferation, but Americans became less anxious about imminent world destruction.

During the early Reagan administration, withdrawal from the Strategic Arms Limitation Talks (SALT II) with the Soviet Union, discussion of prospects for limited nuclear warfare, and proposals to renew civil defense evacuation plans, plus presidential rhetoric characterizing the USSR as the uncontrollable "Evil Empire," fueled public fears of a nuclear holocaust and contributed to an international nuclear freeze movement.[7] As an alternative to nonproliferation, in 1983 President Reagan turned to technology and proposed research on the Strategic Defense Initiative (SDI), a computerized system of ground and space-based interceptors. Also called "Star Wars," this system was intended to sense and destroy attacking missiles. The president represented SDI as a national "insurance policy" against ultimate aggression by the Soviet Union and "a shield that could protect us from nuclear missiles just as a roof protects a family from rain."[8] As the Soviet Union collapsed, this technological defense, widely criticized as ineffective and politically risky, became irrelevant although it was later revived.

During the Clinton administration, officials at the highest level, including the president, publicly expressed their deep fears of bioterrorism, with little evidence that anti-American terrorists were interested in germ weapons. The United States had no single and equal political adversary with which it could negotiate to reduce risks, like the Soviet Union, and it was reluctant to subject itself to multilateral agreements with lesser powers. In this phase, the United States gradually lost its long-range perspective on the importance of states in preventing proliferation, for example, in restraining terrorist acquisition of biological weapons as well as secret state programs.

As it was imagined, bioterrorism reduced the technology of offensive biological weapons to the level of sabotage attacks on urban neighborhoods, sports arenas, malls, or transportation systems, like subways or airports. These limited scenarios became the basis for exercises for local "first responders," the police, firefighters, and paramedics on the local scene. Unlike nuclear, chemical, or large-explosive attacks, the effects of an intentional epidemic would be slow to emerge and last for weeks or longer. Any bioterrorist event would put great demands on public health and hospital resources. The term "dual use" came into new use in reference to public health resources that might be mobilized in domestic preparedness plans. For a while it was hoped that, like the British in 1938, the government would decide that the best defense against biological weapons was a healthy population with good, publicly supported medical care.[9] Instead, the American approach in the 1990s emphasized decentralized civil defense plans and broad technological solutions, such as improved electronic communication and federal antibiotic stockpiles, geared to emergency response.

Biological Weapons Threats from States

Over the years, the list of nations suspected by the United States of having biological weapons programs has emphasized what are often called "niche states," ones that lack the conventional power to challenge the United States, but "possess the resources and know-how to possibly resort to WMD, especially biological or chemical weapons."[10] Cuba and North Korea, for instance, have been listed regularly, along with Iran and, until 2003, Iraq and Libya. On other lists, not all suspects are hostile to the US; some, like Taiwan, have American support. Nor are they necessarily small nations with limited military force. China, Pakistan, and India, each with nuclear weapons, have been cited as perhaps supporting offensive biological weapons programs.[11]

Israel also appears on most lists compiled by analysts and arms control organizations. Little is known about the history of its biological weapons program and its alliance with the United States, which includes military aid, seems to have protected it from American inquiry about its present activities. After the 1948 war, Israel established a biological research unit, called Hemed Beit, which then moved to its permanent location on seventy acres outside the town of Ness Ziona, near Tel Aviv.[12] Known since 1952 as the Israeli Institute of Biological Research (IIBR), the center was sponsored by the Ministry of Defense and staffed by civilian scientists who often had academic appointments and published in scientific journals.

Israel stands apart from other democracies in its refusal to become a party to either the Biological Weapons Convention or the Chemical Weapons Convention. It is a party to the Geneva Protocol, with the reservation that the Protocol would cease to be binding "as regards any enemy State whose armed forces, or the armed forces of whose allies, or the regular or irregular forces, or groups or individuals operating from its territory, fail to respect the prohibitions which are the object of this Protocol." Syria, Jordan, and Libya have long maintained similar reservations specifically allowing retaliation against Israel.

Much more is known about the biological weapons programs of Iraq and apartheid South Africa than any comparable states. Each made different treaty commitments during the Cold War, when their programs were started. Iraq was party to the Geneva Protocol; it signed but did not ratify the Biological Weapons Convention (BWC) until required to do so as part of the 1991 United Nations cease-fire agreement, and it was not a party to the Chemical Weapons Convention (CWC). South Africa was party to the Geneva Protocol and the BWC and, after a regime change, joined the CWC in 1995.

The Threat from Iraq

Before the Gulf War, the lawless characteristics of Saddam Hussein's Iraq were largely overlooked by the United States.[13] Its conventional military force, built up during the Cold War with US and Soviet and other assistance, was a threat to the region. It was known to have used chemical weapons (tear gas, mustard gas, and nerve gas) against Iranian troops, and in 1987 and 1988 it used chemical weapons in attacks that killed Kurdish villagers. Its invasion of Kuwait, to reclaim what it contended were traditional lands, precipitated the 1991 Gulf War, the cease-fire agreement, and the eventual end of Saddam's power.

In 1991, United Nations Security Council Resolution 687 required Iraq to reaffirm its commitment to the 1925 Geneva Protocol, to ratify the BWC, and to accept, under international supervision, the destruction, removal, or rendering harmless of: "(a) all chemical and biological weapons and all stocks of agents and all related subsystems and components and all research, development, support and manufacturing facilities related thereto; (b) all ballistic missiles with a range greater than 150 kilometers and related major parts and repair and production facilities."[14]

UNSCOM, the United Nations Special Commission, chaired by Swedish Ambassador Rolf Ekeus, was assigned to facilitate the elimination of chemical and biological weapons, while the IAEA (International Atomic Energy Agency) was in charge of investigating possible nuclear weapons and succeeded in discovering and destroying Iraq's preliminary efforts in this area. UNSCOM had a double purpose, to disarm Iraq and to continue ongoing monitoring and verification to prevent its reacquisition of the prohibited weapons.[15]

From 1992 to 1995, the UNSCOM team accumulated solid evidence of an Iraqi offensive biological program. Its detailed tracking of purchases of bacteriological growth media by Iraq's Technical and Scientific Materials Import Division showed that thirty-nine tons had been imported in 1988 and several more tons arrived later, with a shelf life of four to five years. This amount was far beyond what Iraqi hospitals, medical laboratories, and pharmaceutical industries could need in that time.

In July 1995, faced with UNSCOM's findings, the Iraqi government was forced to admit it had developed an offensive biological weapons program and to provide information on the amounts of various biological agents it had produced. The Iraqis claimed that all its biological agents had been destroyed and denied ever having placed them in munitions.

In August, Gen. Hussein Kamal, Saddam Hussein's son-in-law and a former top government official, departed for Jordan and there briefly described to Ekeus the development and production of biological agents and asserted that the program had ended. Iraqi officials reacted by arranging for UNSCOM to receive more documentation, at Kamal's farm where, in a chicken house, investigators found a box of written reports, along with microfiches, computer diskettes, videotapes, and photographs of prohibited hardware.[16] Iraqi officials then admitted it had filled munitions with biological agents: five SCUD missile warheads with anthrax, sixteen with botulinum toxin, and four with aflatoxin (which is something of a puzzle, since aflatoxin is a slow-acting carcinogen). They also admitted that bombs and drop-tank aerosol generators for airplanes were part of Iraq's biological weapons arsenal. UNSCOM then intensified its monitoring, making over a hundred site visits, twenty of these to the declared biological production facility at Al Hakam, which the Iraqis, under UN supervision, destroyed in 1996.

According to Iraq's own declaration, as early as 1974, it began exploring the creation of a biological weapons program. Microbiologist Nassir al Hindawi, an anthrax specialist, has been represented as its founding father.

Iraqi scientists and technicians were well educated in Western universities, as undergraduates and graduate students. Rihab Rashid Taha, a microbiologist with a doctorate from the University of East Anglia in the UK, was the main contact with United Nations inspectors.

In 1984 a group of biologists began research within one of the chemical weapons complexes, following the pattern of biological weapons program development in the United Kingdom, the United States, and the Soviet Union. Then, in 1988, Iraq established the Al-Hakam facility for mass production of anthrax, botulinum toxin, and later viruses. With its oil revenues, Iraq had the resources to build or buy the necessary technology, from fermenters to bombs, missiles, and aircraft.

During the 1980s, Iraq was able to purchase four strains of anthrax from the American Type Culture Collection, with the approval of the US Commerce Department. Other pathogens and toxins explored by the superpowers appeared in records of the Iraqi arsenal: tularemia, botulinum toxin, brucellosis, wheat rust, aflatoxin, and ricin. The Iraqis also investigated camel pox, rotavirus, hemorrhagic conjunctivitis, and trichothecene mycotoxin, the alleged "yellow rain" of the 1980s.

According to General Kamal in his August 1995 interview in Jordan, all this activity had ended. Kamal stated that, intimidated by the arrival of UNSCOM, he had ordered the destruction of Iraq's chemical and biological weapons programs in 1991. About biological agents and weapons, his assertion was, "nothing remained."[17]

According to Richard Butler, the Australian ambassador who became the UNSCOM chair in 1997, both Iraq's failure to cooperate fully and internal politics at the United Nations undermined the verification procedure.[18] Russia and France in particular lobbied for approval of Iraqi disarmament efforts and the end to economic sanctions. The resignation in protest of one American inspector, Scott Ritter, raised suspicions that the United States tried to use UNSCOM to further its intelligence agenda, which Butler denied.[19] By December 15, 1998, relations with Iraq had deteriorated and the UNSCOM team was evacuated. Immediately after, the US and UK commenced a punitive four-day bombing raid on Iraq (Operation Desert Fox). The final UNSCOM report on Iraq pointed to missing documentation of compliance, "Iraq has not provided evidence concerning the termination of its offensive BW programme. The evidence collected by the Commission and the absence of information from Iraq, raises serious doubts about Iraq's assertion that the BW programme was truly 'obliterated' in 1991 as it claims."[20]

Despite their varying levels of experience and the obstacles posed by the Iraqis, the UNSCOM teams had been energetic in their site searches.[21] Over nearly eight years, dozens of teams of experts conducted and recorded hundreds of detailed inspections. Until 1995, UNSCOM reported, the Iraqis attempted to conceal the entirety of their biological weapons program and used "outright lying, evasiveness, intimidation, forging of documents, misrepresentation of sites and personnel, the denial of access to individuals and the issue of successive FFCDs [Full, Final, and Complete Disclosures] that were fraudulent."[22] By 1998, though, the "absence of information," not any material evidence of biological weapons, made Iraq suspect.[23]

South Africa

Important information about the South African biological weapons program was acquired after a regime change that brought democracy and the franchise to the majority black African population. The new government voluntarily undertook the public disclosure of the combined chemical and biological weapons program, and its courts commenced legal proceedings that provided more information. UNSCOM and US leaders often compared Iraq's lack of cooperation to South Africa's willing exposure of its past. The South African process was, in fact, less than ideal and much more remains to be known about the program and its activities, both national and across national borders. Although unlike the Japanese program in China, the South African venture was also marshaled against a civilian population considered racially inferior, with the intent of subjugating or exterminating them.

In 1978, P. W. Botha, the former minister of defense, was elected president of South Africa. In the name of national security, he increased military, police, and special operations forces to promote a "total national strategy" against the "total onslaught" of terrorist attacks by the African National Congress and Rhodesian insurgents, and by guerrillas from neighboring states.[24] Soon after, Wouter Basson, a young military physician, was sent overseas to educate himself on chemical and biological weapons.

In 1981 South Africa initiated a combined chemical and biological weapons program in reaction to political threats to its apartheid government and its policies of segregation, which had already distanced it from the Commonwealth and the West. As the Soviet Union, Cuba, and China supported black liberation movements in Angola, Mozambique, Rhodesia (now Zimbabwe), and South West Africa (now Namibia), South Africa identified itself as an anticommunist stronghold and used the South Africa Defense Force (SADF), assisted by chemical and biological weapons program officials, to support the last colonial regimes in Africa.[25] In 1981, Botha appointed Basson the head of the new program, code-named Project Coast.

Although Project Coast was under the South African Defense Department's Surgeon General's office, Basson was subject to little supervision. The project involved South African universities and private industries in covert research projects. It likely had communication about biological weapons with Israel, which shared South Africa's sense of siege and cooperated with it on nuclear and conventional weapons programs. The project's growth years were from 1982 to 1987, when it developed a range of biological agents (such as those for anthrax, cholera, and the Marburg and Ebola viruses and for botulinum toxin) and had plans, probably never met, to build a large, secret production facility.

Counterinsurgency agents were a specialty, from tear gas to sedative drugs to the hallucinogen BZ, which the US had produced and stockpiled in the 1960s and then abandoned. Such "incapacitants" were used to facilitate the killing of hundreds of prisoners from South West Africa, whose bodies (unmarked by overt violence) were then dumped from airplanes into the sea. Project Coast reportedly had research plans, however dubious, for a "black bomb" that would selectively kill or weaken black insurrectionists in troubled, mixed-race areas. The covert sterilization of black people with drugs or vaccines was apparently another project goal.

The end of the Cold War, the 1989 election of F. W. De Klerk to the presidency, and the 1994 political ascendancy of Nelson Mandela and the African National Congress ended the South African chemical and biological weapons program, along with its nuclear one.

In the final years of the apartheid government, British intelligence discovered that Wouter Basson was traveling to Libya as a consultant; he also had made frequent trips to Eastern Europe and Iran and cultivated contacts with racist militia sympathizers in the United States.[26] Basson was the kind of wild-card scientist that US and UK intelligence agencies feared would sell information to hostile states and terrorists. The United States and the United Kingdom urged the Mandela government, which lacked the legal means to restrain Basson, to rehire him.

In 1997, President Mandela asked that the history of Project Coast be considered for hearings before the Truth and Reconciliation Commission. Encumbered by some government restrictions, the hearings proceeded and former program scientists and administrators gave precise, disturbing accounts of their work and its use for sabotage and war.[27] Basson, delaying his appearance to the final hours of the last day, kept his testimony short and evasive.

A subsequent criminal trial of Basson commenced in October 1999 and lasted until April 2001. The judge, who openly approved of Project Coast, finally cleared Basson of all charges. The prosecutors immediately appealed for a new trial with a new judge, which led to years of judicial deliberations, still inconclusive. Nonetheless, both public processes, the Truth and Reconciliation testimonies and Basson's trial, disclosed important information about the secret program.

The Clinton Administration and Terrorism

During the 1990s, widespread fatal epidemics were seriously disrupting sub-Saharan Africa, South Asia, South America, and the Caribbean, especially Haiti, and making inroads in Russia and Eastern Europe.[28] Most people in the United States felt little connection to the risks of disease, poverty, and violence afflicting much of the rest of the world. In the 1980s, the AIDS epidemic had introduced Americans to a devastating emerging infectious disease; with new medical cocktails, HIV infection became more like a chronic disease than a fatal epidemic. Medical technology had contained the threat of collective death. If there was a worry, it was that federal withdrawal from medical and social welfare programs would leave Americans prey to for-profit hospitals, managed care, and drug companies.

In this context, terrorist attacks of increased scale and daring opened the door to thinking that biological weapons attacks would be next, if not nuclear weapons or nerve gas. The 1993 bombing at the New York World Trade Center, planned and executed by international terrorist Ramzi Yousef, was the first event that demonstrated how easy foreign attack might be. Only chance prevented it from having its planned catastrophic impact, that one of the Twin Towers would topple against the other.[29]

In 1995, the Oklahoma City bombing again demonstrated US vulnerability, this time from within, from an American, Timothy McVeigh, a Gulf War veteran inspired by militia extremism and the Christian Identity movement. The violence of the event shocked the nation with its scale of destruction and death (168 killed, including 19 children). In response, President Clinton issued Presidential Decision Directive (PDD-39), which laid out an agency-by-agency national strategy for preventing and responding to terrorist attacks.[30] PDD-39 was the first high-level, major policy document to address the problem of large-scale terrorism with centralized federal controls. The increase in the federal counterterrorism budget went from $5.7 billion in 1995 to $11.1 billion in 2000.[31]

The US Congress reacted with bipartisan legislation to bolster preparedness against WMD terrorism in state and local communities. Senators Sam

Nunn and Richard Lugar, who had sponsored legislation to reduce the post-Soviet threat, were joined by Senator Pete Domenici in sponsoring the 1997 National Defense Organization Act, which established the Domestic Preparedness Program. Although the United States was not at war, the theme of decentralized civil defense was popular in Congress; it meant that funds would flow to cities and states to reinforce emergency police, firefighter, and medical resources. The Nunn-Lugar-Domenici legislation gave the Defense Department the lead in allocating grants to the nation's 120 largest cities for training, equipment, and exercises to prepare them for WMD attacks.

Two minor links between militia extremists and the possession of dangerous pathogens surfaced in the 1990s. One was the 1994 case of the Minnesota Patriots Council, in which four members derived ricin from castor beans with the intent to kill government officials.[32] The other was the case of Larry Wayne Harris, arrested in 1995 and again in 1998 on suspicion of planning to create a biological weapon using anthrax spores. Harris, who received extensive media coverage, was released both times.[33] The Minnesota militia members were sentenced to jail in 1994 and 1995 for violation of the 1989 Biological Weapons Antiterrorism Act, which extended the prohibitions in the BWC (including possession) to federal law. Both these cases, along with Iraq's earlier purchase of anthrax strains, sounded the alarm about the commercial availability of dangerous pathogens. Still, foreign terrorists appeared a much more substantial threat of WMD attacks than America militias.

Foreign Terrorists and WMD

A month before the Oklahoma City bombing, members of the Aum Shinrikyo (Shining Truth) cult released nerve gas in an attack in the Tokyo subway.[34] The delivery system was little more than multiple plastic bags with holes poked in them by umbrellas. In addition to the twelve people who were killed, a thousand were hospitalized and thousands more sought care. Eight months earlier, in the town of Matsumoto, the cult had released a sarin cloud that killed seven. The event, although reported in the press, went largely unnoticed in the United States. Further investigation revealed that the cult had experimented unsuccessfully in Japan with dispersing anthrax spores in public places. It also had a New York City office and plans to attack there and in Washington, DC.

The FBI had already had a brush with a cult group's hostile use of a pathogen. In 1984, the Oregon members of the Indian Rajneeshee cult covertly sickened 751 people by contaminating ten local restaurant salad bars and coffee stations with *Salmonella typhimurium* purchased from a US medical supply company.[35] The following year, after the cult had dispersed, a former member voluntarily confessed the plot, saying it was part of a plan to keep local voters from impeding Rajneeshee expansion. Until the confession, the outbreak was attributed to poor restaurant sanitation.

Aum Shinrikyo's organization was global. In post-Soviet Russia, it recruited thirty thousand members and had twenty thousand or more in Australia, Germany, Taiwan, and the former Yugoslavia. Its war chest, conservatively estimated at $20 million, reflected its support by members. Unlike other active terrorist organizations, Aum Shinrikyo attracted scientists and explored the possibilities of other lethal agents and perhaps nuclear weapons.[36] Its goal was apocalyptic, the destruction of the corrupted world order in order to create a new, pure society.

Much of the cult's success in attacks resulted from the Japanese government's reluctance to intrude on any of its many religious sects. Japan also had to create new laws to prosecute individuals for WMD possession and use. Nine of the Aum's members were given death sentences by the Japanese courts in a slow legal process that kept Asahara Shoko, the group's leader, in jail for nearly ten years before he received the same sentence. Aum Shinrikyo and its chemical and biological activities strongly influenced how the US government perceived the global terrorism threat: apocalyptic, international, equipped with the financial assets and scientific skills to develop and use weapons of mass destruction.

Postmodern Terrorism and the Middle East

Since the 1980s, US troops and embassies in the Middle East and Africa had been the targets of terrorist attacks. The globalization of Islamic extremist terrorism was a new phenomenon, epitomized by Al Qaeda (the Base) and its leader, Osama bin Laden, a son of a Saudi millionaire. Bin Laden was a long-time sponsor of Sunni Islamic extremism in Afghanistan, where during the 1980s the United States had supported his guerrillas against Soviet troops.[37]

In August 1998, Al Qaeda operatives coordinated the bomb attacks on the US embassies in Dar es Salaam, Tanzania, and Nairobi, Kenya. In the first explosion, eleven were killed and seventy-four injured. In the second, in a congested downtown area, 213 were killed and 4,500 were treated for injuries. The Al Qaeda coordination of the embassy bombings, which involved terrorists from Egypt, Jordan, Saudi Arabia, a US citizen, and several Africans, supported bin Laden's claim of control over an international network united in an Islamic jihad against Israel and against American military "Crusaders" in the Middle East.[38] As Secretary of State Madeleine Albright commented in reaction to the bombings, they seemed to indicate "a clash between civilization itself and anarchy—between the rule of law and no rules at all."[39]

The Aum Shinrikyo attacks and the Al Qaeda bombings lent credence to theories that terrorism had entered a new postmodern age of violence, driven by religious fanaticism.[40] Or, as some hypothesized, the conflict between the Western civilization and fundamentalist Islam would define future wars.[41]

In reaction to the embassy bombings, President Clinton responded with military force against two targets.[42] Following a CIA report of a chemical precursor (called EMPTA) to the nerve agent VX found near a Sudanese factory, Clinton ordered the factory bombed. The president also ordered the bombing of six terrorist training camps in Afghanistan, where around sixty people were killed. Critics of these strikes and of the December 1998 Desert Fox attacks on Iraq suggested that President Clinton, beset by a perjury and sex scandal, was creating a political diversion. Others felt he did not go far enough in attacking the terrorist camps.[43] Clinton's resort to punitive military force to counter terrorism had a precedent. Earlier, in 1994, President Clinton had authorized cruise missiles against an Iraqi intelligence center in retaliation for its attempt to assassinate former President Bush while he was in Kuwait. Even before that, in 1986, President Reagan had ordered an air strike against Libya in retaliation for the terrorist bombing of a West Berlin disco frequented by US soldiers. Like Reagan's air strike, Clinton's air attacks were intended as a warning blow to a lesser nation that might encourage anti-American terrorism. That Clinton relied on a scientific test, later disputed, to justify bombing Sudan was the unusual aspect of that decision.[44]

In the aftermath of attacks on the United States abroad and at home, "a permanent crisis mode" dominated the White House. At the National Security Council Richard Clarke made Al Qaeda his principal target and worked on strategies to destroy the network with the CIA, Defense Department, Justice Department, and the FBI.[45]

Science Advising and Visions of the Apocalypse

The first high-level Clinton appointee to promote biological weapons as a national security threat was Secretary of the Navy Richard Danzig.[46] Danzig, a Yale Law School graduate like Clinton, closely followed the Aum Shinrikyo case and considered it a likely model for future WMD terrorism. He was optimistic about Defense Department programs for a range of technologies (from biosensors to test air to improved vaccines) and for defensive troop training and the development of doctrine to preempt and respond to biological attack.

In 1997 President Clinton, concerned about the possible terrorist use of genetically altered biological agents, turned to microbiologists for advice. Like Danzig, Clinton relied on Nobel laureate Joshua Lederberg. For decades Lederberg had consulted for US intelligence and defense agencies.[47] In addition, he had long warned of the possibility of devastating epidemics of exotic infectious diseases. In 1968, in reaction to a small epidemic of just such a disease (Marburg fever), he wrote in the *Washington Post* of the "global pandemic ... that hangs over the head of the species at any time."[48] In 1997, in the same apocalyptic mode, Lederberg adjusted his warning to the times: "Visualize the World Trade Center [in 1993] or an Oklahoma City–style attack complicated by the inclusion of a kilogram of anthrax spores as a kind of microbiological shrapnel along with the explosives. And [imagine] its implications for salvage and rescue, public health, panic. If I just mention the word Ebola, you have some idea of what I am talking about."[49]

Another science advisor to the president was Craig Venter, founder of the Institute for Genomic Research and of Celera, the company that first sequenced the human genome. Venter urged sequencing the genomes of dangerous pathogens as a possible way to design improved measures against them.[50] Like Venter, Lederberg advocated a technological solution, the use of basic research to develop new drugs and vaccines, which could also protect against emerging infectious diseases.[51]

In November 1997, Secretary of Defense William Cohen appeared on national television to make the viewing public conscious of an imminent mass biological weapons threat. Holding up a five-pound bag of sugar, Cohen stated that an equivalent amount of anthrax, if sprayed by plane over Washington, DC, would kill half its population. Having articulated the possibility of this horrible attack, Cohen could offer for protection the Nunn-Lugar-Domenici bill, the decentralized domestic preparedness program the Defense Department would administer.

The other part of Cohen's speech announced a pharmaceutical solution to protect US troops abroad. Through a new Anthrax Vaccine Inoculation Program (AVIP), all 2.3 million American military personnel would be vaccinated against the risk of a biological weapons attack by a hostile nation such as Iraq or by a foreign terrorist group. This vaccination campaign was Secretary Danzig's idea, and he had enlisted Lederberg to overcome the reluctance of Pentagon leaders.[52] Anthrax vaccinations were already standard for soldiers about to be deployed to high-risk areas, such as the Middle East or the Korean Peninsula. This new program was tied to federal support for a private company that was to be the army's sole supplier. The former head of the Joint Chiefs of Staff, Adm. William Crowe, sat on its board and had testified before Congress on behalf of the venture.

To advertise the vulnerability of the public to mass biological attack was in itself a risky approach, especially when few safeguards were in place. Who knew what ideas such an announcement would stimulate? The Cohen speech was followed by a precipitous rise in anthrax hoaxes, with hundreds of envelopes of sugar or other white powders posted to individuals, schools, offices, and churches and causing anxiety and disruption.[53] Another result was media exploitation of fears of "intentional epidemics" on television, in films, and in the press, with political and medical authorities warning the public that a bioterrorist attack was not a matter of if but when.

In 1999, after several years of low funding, the annual budget for WMD domestic preparedness programs reached $10 billion. Seventy percent was allocated to the Department of Defense and most of the rest went to the Justice Department and the Department of State.

Although Secretary Danzig called for an "enhanced cooperative relationship" between the military and federal agencies charged with civil defense against biological weapons, the scope of that cooperation remained undefined. In 1994, Clinton had supported the National Defense Authorization Act, which gave the Defense Department the responsibility for preparation and response to domestic nuclear, chemical, or biological attacks. In the remote event of an enormous catastrophe, such as a nuclear strike, the military was prepared to evacuate survivors. Its role in response to chemical or biological attack was uncertain. Afterward, Congress funded a dozen of what Cohen dubbed RAID (Rapid Assessment and Initial Detection) National Guard units, for integration into local response plans, which proved unsuccessful in practice. In mid-1998, the Pentagon proposed establishing a national military command to coordinate a National Guard and other military responses to emergencies. It then retreated when the American Civil Liberties Union and influential members of the press protested. They argued that, if crossed, the line between military and police enforcement, established after the Civil War with the Posse Comitatus Act, could lead to violations of civil rights.[54]

In 1998, in a new directive (PDD-62), President Clinton further delineated the federal counterterrorism responsibilities and placed them squarely within civilian agencies. Within the Department of Justice, the FBI would handle "crisis management": frustrate or stop a terrorist attack, arrest terrorists, and gather evidence for criminal prosecution. FEMA (the Federal Emergency Management Agency) would be in charge of "consequence management": the provision of medical treatment, evacuation of populations in danger, and the restoration of government services. These two task distinctions would later be reformulated as the number of federal agencies and offices potentially mobilized by a grand-scale terrorist attack increased to more than forty.

Simulations of Civil Biodefense

In major cities across the country, the federal government helped stage simulated biological, chemical, and radiological ("dirty bomb") attacks to mobilize local officials for emergency response. Police, firefighters, and emergency medical teams rehearsed the rescue of afflicted civilians, played by community volunteers. "Domestic preparedness" was far from standard. Local government leadership determined the degree of participation, the plans for mobilization, and what resources would be requested from the federal government—whether new computers or police cars or ambulances or support for personnel and training.[55]

In 2000, to test if domestic preparedness legislation was improving national readiness, Congress asked the Justice Department and FEMA, together with the National Security Council, to stage an exercise that would mobilize top government officials in a simulation of an attack response. The exercise, called TOPOFF (for "top officials"), was directed by an established defense contractor, SAIC (Science Application International Corporation). Costing around $10 million, TOPOFF simulated a mustard gas attack in Portsmouth, New Hampshire, where the response went smoothly, and a scripted plague aerosol attack in Denver, Colorado. The Denver exercise produced chaos (locating mortuaries to store fictive cadavers was a major problem) and brought out the difference between a limited chemical attack and one with a contagious disease. This difference was important. An explosion or chemical attack would be immediately evident and localized. A disease outbreak, though, could be undetected at first and then last over weeks. Patients could leave the attack locale without realizing they were infected, fail to understand the gravity of their illness, and, if the disease were contagious, perhaps spread and prolong the epidemic.

In July 2001, another bioterrorist simulation, called "Dark Winter," emphasized the contagious disease threat, even more strongly, on the scale of war. Held at Andrews Air Force Base, Dark Winter was a tabletop exercise, based on a fictional pandemic of smallpox; the scenario condensed the events of thirteen days into two.[56] The invited participants or actors were Washington political insiders. For example, Senator Sam Nunn, a key sponsor of domestic preparedness legislation, played the part of the president. The script, written largely by staff at the Johns Hopkins Center for Civilian Biodefense Studies, illustrated a worst-case scenario in which smallpox spread across the nation, where insufficient vaccine was available to stop it. The US military had to intervene to curtail violence and social breakdown. Then a worldwide pandemic was added.

The Dark Winter scenario was later criticized by scientists at the Centers for Disease Control (CDC) and other infectious disease experts for its exaggerated contagion rates and its lack of emphasis on proven simple ways to curtail epidemics, such as home care, wearing face masks, hand washing, and, perhaps most important, avoiding hospitals where transmission rates would soar.[57] In fact, the exercise served well as political rhetoric. Two weeks later, its organizers and participants testified before Congress in support of increased funding for stockpiling smallpox vaccine and for domestic response training.[58] The exercise showed that the scale of an imagined bioterrorist attack could vary greatly, according to the scriptwriters and their intents.

During 2002, as polls indicated, the American public became more intimidated by the possibility of a nationwide smallpox outbreak.[59] Among experts, concern about a future smallpox outbreak with perhaps a new strain had a direct impact on indefinitely delaying the WHO scheduled date (December 31, 2002) for destroying the US and Russian reserves of the virus, the last known in the world.[60] Following the WHO smallpox eradication campaign, no case of the disease had been recorded since 1979. Experts who saw basic science as the key to defense against bioterrorism envisioned the development of antiviral drugs to replace current vaccines, which, although valuable, were already contraindicated for people with compromised immune systems. Political justification for the delay was found in fears that Saddam Hussein might use smallpox in a last-stand attack or that North Korea's Kim Jong-il would do the same.[61] By this reckoning of future threats, it could be argued that the smallpox stocks should be preserved for research purposes.

Donald A. Henderson, a leader of the WHO smallpox eradication campaign and founder of the Johns Hopkins unit that organized Dark Winter, disagreed. His solution was for the government to destroy the virus and stockpile enough smallpox vaccine to counteract an American pandemic.[62] The smallpox virus itself is not used in making the present vaccine; better vaccines, Henderson argued, could be developed without retaining the virus. A report from the Institute of Medicine disputed the wisdom of destroying the virus.[63] In agreement, the WHO delayed the extinction of smallpox. Developing nations most vulnerable to smallpox reemergence protested. At the same time, the US government moved forward with the production and stockpiling of the smallpox vaccine in the event of a bioterrorist attack on America.

Public Health and Bioterrorism

If bioterrorism posed a collective infectious disease threat, public health seemed the obvious response. The United States, though, had only inconsistently supported public health, which was often seen as a way in which government might curtail individual liberties and the operation of a free market.[64] In the 1990s, the low-status and underfunded American public health system was charged with disease prevention and health care for the disadvantaged, such as AIDS and hepatitis testing, prenatal care, childhood vaccinations, drug abuse prevention, annual influenza shots for the elderly, and the laboratory monitoring of disease outbreaks. Its professional organizations and schools were also oriented to international infectious disease problems; national security was a limited and even unrealistic framework for risk reduction in a world with accelerated global travel, trade, and movement of populations.

In 1999, the Clinton administration heralded the integration of public health and national security to fight the threat of bioterrorism.[65] But this integration was oriented not toward a reinforcement of, say, Medicaid, the nation's most comprehensive public health program, but toward technological solutions such as electronic disease surveillance and reporting, better medical diagnostic tests, and improved surveillance of water supplies and food production.

Public health physicians became concerned about the impact that "civilian biodefense" and its emphasis on emergency response could have on their role in providing routine services and protecting patients' rights. Victor Sidel, the public health leader who advocated the elimination of the US biological weapons program, saw a conflict between what were fundamentally national security goals and professional responsibilities to patients. He made the point that "military, intelligence, and law enforcement agencies and personnel have long histories of secrecy and deception that are contrary to the fundamental health principles of transparency and truthfulness. They may therefore be unsuitable partners for public health agencies that need to justify receiving the public's trust."[66]

Troubling issues of military deception and secrecy had already tainted the Department of Defense's universal anthrax vaccination program, AVIP. Side effects were being underreported or suppressed by the Pentagon. The private pharmaceutical company had failed to win FDA approval for vaccine production and was relying on old, faulty stocks inherited from the last manufacturer, the State of Michigan.[67] Soldiers refusing the vaccination were being dishonorably discharged. Much of this information was released only because independent critics, such as Dr. Meryl Nass and Victor Sidel, and the families of soldiers had pressured Congress for investigation.

Sidel and others who emphasized trust and openness in disease management spoke from practical experience. In all outbreaks, accurate information—about the source of the disease, its nature and transmission, about who might have been exposed and why and where, and about how the victims can be quickly helped—is crucial to local public awareness and mobilization and, on clinical level, to early diagnosis and saving lives.[68] The 1979 Sverdlovsk outbreak provided a worst-case example on a small scale of how military and government secrecy can have deadly consequences for the public. A larger outbreak similarly fraught with misinformation and disinformation would be a true catastrophe.

The millennium ended without the predicted bioterrorism event. The perceived threat of biological weapons, rather than diminishing after the end of the Cold War, continued to increase.

CHAPTER 9

National Security and the Biological Weapons Threat

The September 11, 2001, Al Qaeda attacks on the United States and the anthrax postal attacks that soon followed together intensified Clinton-era policies already in place for national security and defense against biological weapons. The federal intelligence budget rose again. The domestic preparedness programs in place throughout the United States received more support, as did funding for technological defenses, from air sensors to new pharmaceuticals. The great change after 9/11 was in the organizational scale of the Bush administration responses to this unprecedented foreign attack on Americans. The minor use of force, that is, the 1997 bombings of Afghanistan and Sudan, was overshadowed by military invasion in the name of the war against terror, whether the threat was training camps or Iraq's weapons of mass destruction. The 2003 creation of a Department of Homeland Security far outweighed the diffuse, decentralized domestic preparedness project of the previous decade. Clinton's fortification of a few public health technologies was dwarfed by sweeping policies to incorporate US biological sciences into the campaign against bioterrorism.

Each of these expanded policy initiatives brought dilemmas concerning the flow of information and public trust in government. The invasion of Iraq was eventually followed by disclosures suggesting that the Bush administration manipulated intelligence data in order to convince Americans of the imminent threat of Saddam Hussein. The violent and costly military occupation that ensued further troubled public trust in the administration. The Department of Homeland Security from the very start faced the problem of relying on intelligence information to pronounce public alerts that were credible, even when no crisis occurred. Could a state of emergency be indefinitely maintained? In addition, there was the question of whether local first responders would be in the Washington information loop or, as happened during the anthrax letter attacks, would physicians and targeted victims be the last to know about the risks? Last, the large-scale government recruitment of biologists to the biodefense campaign signaled new rules for secrecy in basic scientific research that, until the present, has been open in its beneficent intents and goals.

Secrecy and Defensive Research

Even before September 11, the presidency of George W. Bush signaled a return to Reagan-era policies, which emphasized use of force and technology for national defense, less as a complement to international agreements than as a substitute for them. In charge of the world's mightiest military nation, the Bush administration proved much more disdainful of arms-control treaties than President Reagan or the former President Bush. In July 2001, it withdrew from negotiations for a legally binding verification and compliance Protocol to the Biological Weapons Convention. It stood opposed to ratification of the Comprehensive Nuclear Test Ban Treaty, and in July 2002, it withdrew the United States from the 1972 Anti-Ballistic Missile Treaty. These decisions and the 2003 unprovoked invasion of Iraq contributed to the perception abroad that the United States claimed exception from international norms against armed aggression.

From the beginning, the Bush administration took the position that military secrecy and pharmaceutical commercial interests took precedence over negotiations to strengthen the Biological Weapons Convention (BWC) with mandatory compliance measures. In December 2001, having rejected negotiations, the Bush administration put pressure on the Ad Hoc Committee deliberating the protocol to disband entirely. In a compromise, Hungarian ambassador Tibor Tóth, chair of the committee, agreed to a three-year plan for continued discussions on implementing the treaty, with verification off the agenda. The issues for discussion became limited to subjects that did not trouble military or commercial interests and that states could pursue independently of a general accord. These agenda items were national enactment of penal legislation; national oversight of dangerous pathogens; measures for investigating and responding to allegations and suspicious disease outbreaks; the surveillance and combating of infectious diseases of humans, animals, and plants; and the development of codes of conduct for scientists.

Washington's resistance to transparency was better understood when, in September 2001, the book *Germs,* written by three investigative journalists, revealed secret US projects that blurred the line between defensive and offensive development.[1] These projects included the making and testing of a model of a Soviet anthrax bomb, the building of a small biological-weapons production facility from material bought on the open market, and the creation of a vaccine-resistant strain of anthrax. They were kept so secret that the first two projects were unknown to the Clinton National Security Council staff and to the president himself until they were tipped off by the journalists, who at the time had yet to reveal the information to the public.

For the bomb project, the CIA consulted with scientists, among them Joshua Lederberg, who suggested a legal review regarding possible violation of the BWC prohibitions. CIA lawyers decided the project was within the allowed realm of defensive research. CIA lawyers decided the same for the second project, on the small, fully functional biological weapons facility, sponsored by the Defense Threat Reduction Agency (DTRA) within the Defense Department. The third project, developing a vaccine-resistant strain of anthrax, was meant to duplicate a Russian experiment done in the early 1990s and published in the open literature.[2] When the CIA proposed that the United States make its own version, the Clinton administration hesitated. In the first months of the Bush administration, National Security Council officials gave their permission. They believed the Pentagon had the right to investigate genetically altered pathogens in the name of biodefense, "to protect American lives."[3] The Pentagon then authorized the Defense Intelligence Agency (DIA), under its secret "Project Jefferson" program for pathogen research, to re-create the strain.

The news of these projects confirmed the administration's exceptionalist national security position. They showed that the United States military and intelligence agencies had the power to stretch the limits of the treaty without public review and beyond what would be tolerated for other countries.[4]

Iraq and Weapons of Mass Destruction

After September 11, the members of the UN Security Council and the international community generally supported the US invasion of Afghanistan later that year and the destruction of the Taliban government. All of the major powers had experienced terrorist violence against civilians and their sympathy for American losses from the Al Qaeda attacks ran high. The Security Council, however, became deeply divided over the US planned invasion of Iraq in 2003.

For some, this aggression indicated a dangerous disregard for international legal restraints.[5] But no nation or alliance of nations was in a position to countermand the US decision and the United Kingdom sided with the Bush administration. At the highest level of US policy making, the decision had already been made to topple the government of Saddam Hussein.[6]

In their appeals for public support for an invasion, President Bush and British Prime Minister Tony Blair portrayed the imminent danger of Saddam's weapons of mass destruction as the primary justification for war. In coordination with a speech of Blair's to Parliament, his office released an intelligence summary to the London *Evening Standard*, which, on September 24, 2002, headlined that Saddam was "45 minutes from attack" with chemical weapons capable of reaching British military bases in Cyprus.[7] Meanwhile, France, Russia, and Germany were arguing for continued containment of Iraq and more time for UN inspectors to complete their mission. To forestall an invasion, on November 8, 2002, the UN Security Council adopted Resolution 1441, requiring immediate, unconditional, and active cooperation from Iraq. Shortly thereafter, UNMOVIC (United Nations Monitoring, Verification and Inspection Commission), the successor to UNSCOM, began onsite inspections in Iraq.

The Bush administration's resolve to destroy Saddam Hussein's regime may well have been set years before, completely apart from 9/11. In 1991, George W. Bush's Vice President, Richard Cheney, had been Secretary of Defense for President George H. W. Bush. At that time he articulated an aggressive containment policy for dealing with rogue states, whereby they should expect massive military retaliation for their aggression and a bombing of their weapons of mass destruction "out of existence."[8] The cease-fire with Iraq halted the Gulf War before this policy could be carried out, but Cheney had a new opportunity to follow through in 2001.

The series of inspection updates from UNMOVIC failed to persuade the US government that Iraq was no longer an immediate, serious threat. On January 27, 2003, Hans Blix, the executive director of UNMOVIC, reported to the United Nations his evaluation of progress thus far.[9] He said, "Iraq has on the whole cooperated rather well.... The most important point to make is that access has been provided to all sites we have wanted to inspect." UNMOVIC's one hundred trained inspectors had made three hundred inspections, twenty to sites never before visited. It had recently acquired the use of eight helicopters that helped speed inspection. Germany had promised assistance with unmanned surveillance vehicles to monitor changes. Blix also described incidents when Iraqi officials were uncooperative, although they seemed more willing to assist as pressure mounted. About biological weapons, Blix reported that nothing had been found but that a store of anthrax "might still exist" and "should be found and be destroyed under UNMOVIC supervision or else convincing evidence should be produced to show that it was, indeed, destroyed in 1991."

In early February, Secretary of State Colin Powell made a ninety-minute presentation at the United Nations that, complete with slides and audiotapes, included US intelligence data on Iraq's programs. Special emphasis was put on mobile facilities for producing germ weapons and photographs were shown of trucks, vans, and boxcars allegedly involved in this venture. Holding up a small vial of anthrax simulant, Powell reminded his audience that this same amount, "a few teaspoons," in the 2001 anthrax letters had killed five people and caused great fear and disruption.

Powell failed to say that UNMOVIC investigators had methodically followed through on US intelligence and found no evidence of chemical or biological weapons at suspected sites; nor had they found evidence of mobile facilities. Shortly after, Blix again reported to the UN and raised questions about faulty US and UK intelligence. At the same time, Mohamed Elbaradei, director general of IAEA (the International Atomic Energy Agency), reported that the agency found no evidence of ongoing nuclear or nuclear-related activities in Iraq; his organization had supervised the removal of nuclear materials in 1995 and, under containment policies, no further production had occurred. Powell's assertion that it was a "fact" that Iraq had a nuclear program left the erroneous inference that Iraq's nuclear threat was imminent. This assertion was much repeated by officials in the Bush administration.[10] Still, it remained that the Iraqis under Saddam Hussein had failed to be fully cooperative.

Disregarding the reports of UN inspectors, the United States, along with UK forces, invaded Iraq on March 19, 2003. That May, President Bush declared the cessation of major military operations. For many months after, no weapons of mass destruction or long-range missiles were discovered in Iraq. In January 2004, David Kay, a former IAEA investigator in charge of the US Iraq Survey Group, concluded after six months of searching that Saddam's WMD arsenal had been destroyed in the 1990s, during the UNSCOM investigation.[11]

In the United Kingdom, criticisms of Prime Minister Tony Blair for exaggerating the Iraq WMD threat were severe, and two members of the Blair cabinet resigned in protest over the issue. By June 2003, news headlines were dominated by accusations that Blair had manipulated intelligence information to justify British participation in the invasion. In July 2003, respected former UN inspector David Kelly, a consultant for the UK Ministry of Defense as well as the Foreign Office, committed suicide after being publicly embroiled in the controversy. The national commission, headed by Lord Hutton, that investigated the circumstances of Kelly's death generated even more headlines, although it found no evidence of foul play and avoided blaming Blair's staff for deliberately misinterpreting intelligence data.

Around this same time in the United States, Bush administration's use of intelligence information was only mildly criticized in the press. In his January 2003 State of the Union address, President Bush had claimed that Iraq had sought uranium from an African state. In July, news leaked that the CIA had not authorized that claim; rather, one of the National Security Council staff had inserted it in the speech.[12] Another administration claim, that Iraq had purchased special aluminum tubes for nuclear-weapons production also turned out to be false. The Bush administration's restating of why the war was necessary—because Saddam Hussein was a ruthless dictator who had used weapons of mass destruction and might do so again—was generally accepted as sufficient justification for a war that seemed to have ended victoriously.

The Anthrax Letters and Government Response

During the 1990s, the US Government Accounting Office and various critics pointed out that domestic preparedness programs were inexcusably uncoordinated and would be ineffective in the event of WMD terrorist attack.[13] In the autumn of 2001, government responses to the anthrax letters provided an unpredicted real-life demonstration of organizational problems and their repercussions for victims.

The anthrax postal attacks blindsided government agencies, which in their responses pursued sharply different goals. The Centers for Disease Control had an epidemiological mission as well as the responsibility to contain the outbreak. The FBI was intent on a criminal investigation. Military and intelligence agencies had roles to play, along with the US Postal Service, the Environmental Protection Agency, and a host of other federal and local responders. The flow of information was frequently blocked, within and across bureaucratic divisions. At the height of the crisis, from mid- to late October, as many as three federal press conferences were held a day, each conveying information that was often confused and confusing.[14]

The anthrax letters were likely mailed in two batches from letterboxes in Princeton, New Jersey. The first set of perhaps five letters was posted around September 18 to mass media offices, one to American Media Incorporated (AMI), a tabloid publisher in Boca Raton, Florida, and the rest to broadcasters and editors in New York City. All these early letters were dismissed as routine hate mail or the letters were left unopened. On October 3, 2001, as soon as the first victim, Robert Stevens, was diagnosed in Florida, teams from the Centers for Disease Control in Atlanta were dispatched to investigate the source of Stevens's infection. Their first hypothesis was that he had caught the disease from spores in soil, with little attention paid to the possibility of exposure to a biological agent from any source. When the CDC discovered anthrax spores at Stevens's workplace, the AMI building, the FBI assumed authority and the disease event was framed as a criminal investigation. Evidence emerged that Al Qaeda operatives might have been in the area, not far from AMI. The idea that the source could be a US government-connected scientist only gradually took hold, at least in the public mind.

For the CDC, anthrax was a rarity that posed no public health hazard. The FBI was familiar with limited aspects of anthrax spores as select agents. Since Secretary of Defense Cohen's 1997 speech on anthrax, the agency had been notified of hundreds of hoax letters containing sugar or other powders. The most comprehensive knowledge of anthrax had been developed by the US offensive program, but most of those institutional memories were lost. Before the anthrax letters, the CDC and the FBI were unlikely to share information, although the FBI and Fort Detrick staff did communicate about potential bioterrorist threats. In addition, the postal attacks were such a novelty that only a few officials imagined how mailed spores might be dangerously dispersed.

In 2000, after hoax anthrax letters were mailed to the Canadian Parliament, scientists from the Defense Research Establishment at Suffield (DRES) used *B. globigii* (BG) spores in experiments on how anthrax spores might disperse from an envelope opened in an ordinary office setting. The DRES scientists found that even a tenth of a gram of BG dispersed into the air throughout the office space far beyond all previous predictions, and that a person breathing the air would have inhaled spores in numbers thousands of times the estimated human LD50 (the lethal dose for half an exposed population).[15] Such a letter also might, the report suggested, leak spores from unsealed edges. This initially classified information was shared with the FBI and US military but the CDC was unaware of it. On hearing about the first Florida case of anthrax, the DRES scientists emailed the report to the CDC, where officials were too understaffed to read it until the anthrax postal crisis was nearly over.

The DRES report speculated about spores leaking from the unsealed corners of an envelope. No one imagined or tested the possible effect of mechanized postal sorting machines or whether anthrax spores, a micron in diameter, would be able to escape through envelope paper, which commonly has imperceptible openings as large as three microns. Few officials knew about the air turbulence that characterizes postal facilities. Without this knowledge, the calculation of risks to postal workers was greatly impeded.

An even greater hindrance to assessing risks from the anthrax letters was an unnecessary confusion about dose response. Officials at the Department of Health and Human Services and within it, the CDC, had little knowledge of the quest of the old biological weapons programs and their ways of creating anthrax aerosols with spores in the optimal one- to seven-micron range. During the crisis, retired Fort Detrick scientists who had researched this problem in detail were not consulted.[16] They could have told CDC officials that, over the years, the Army routinely used an LD50 of eight thousand to ten thousand inhaled spores to estimate standard doses on which to base munitions fill. This extrapolation from animal research was never intended to convey a threshold below which an exposed individual was safe. Although very unlikely, even a single anthrax spore or a small number deposited in the lungs might successfully germinate and cause infection and death; estimates of inhalation per person in the ceramics factory affected by the 1979 Sverdlovsk release, based on US Army methods of calculation, were as low as nine spores per victim.[17] Yet, as if an environment with low levels of anthrax contamination posed no jeopardy, government officials and the media repeated eight thousand to ten thousand spores as a threshold range. And how would a low level of contamination be measured? Neither the CDC nor the FBI was practiced with environmental sampling techniques on which to gauge the risks of anthrax.

A success in the federal response to the anthrax letters was the mobilization of the National Laboratory Response Network. Established in 1999 the network includes one hundred public health laboratories coordinated with the US Army Medical Research Institute of Infectious Diseases (USAMRIID) and CDC. This network absorbed much of the burden of analyzing anthrax hoax materials that flooded local agencies and the FBI by the thousands during the postal attack crisis.

In responding to the anthrax postal attacks, the CDC also showed that the emergency distribution of pharmaceuticals and simple tests for exposure could be accomplished with relative efficiency. Throughout the postal anthrax crisis, CDC and local health officials were able to draw on supplies from the national stockpile established in the Clinton administration. Some 33,000 people received postexposure drugs, usually Ciprofloxacin (often referred to as Cipro) or Doxycycline. It is difficult to say whether all these people were at risk of exposure, but some were surely protected from infection and possibly death. The problem was in the timing of the distribution of antibiotics, which had to be early, before the anthrax bacteria produced their deadly toxins. And this timing depended on an alert and informed public.

In Washington

It was not until October 12, 2001, a week after Robert Stevens died, that it was certain that letters were the source of multiple anthrax infections. By that time dozens if not hundreds of people had been exposed. On October 12, an anthrax-laden letter addressed to NBC news anchor Tom Brokaw was discovered in New York. After a general search of media offices, involving both CDC and the FBI officials, another letter, this one unsealed, was discovered at the offices of the *New York Post.* Two cutaneous cases among New York media workers, initially misdiagnosed, were then identified; the infant son of an ABC news producer also contracted severe, misdiagnosed cutaneous anthrax, following a single visit with his mother to the news office in September.

Before the Brokaw letter was found, the criminal who sent the first set of letters may have been frustrated that none had been discovered and reported on by the media. A second set of anthrax letters was mailed around October 8 from Princeton. Two went to Democratic senators in Washington, D.C. One of these was opened on Monday morning, October 15, in the Hart Senate Office Building office of Tom Daschle, then the Senate majority leader. Within the hour, Capitol Hill police had identified the powder as anthrax and staff from the Office of the Attending Physician began distributing Ciprofloxacine to around forty Senate employees at high risk of exposure, with advice to take the pills for sixty days. In the following two days, antibiotics were distributed to hundreds of other Senate and government workers at the Hart Building, which was more contaminated than officials first realized.

The spores (less than a gram) remaining from the Daschle letter were immediately transported to Fort Detrick and analyzed. The next day, in a press conference, Senator Daschle characterized the anthrax letter material as "weapons-grade," the description just used in a closed interagency briefing for Congressional leaders. From the White House Homeland Security office, former Pennsylvania governor Tom Ridge, its head, objected to this characterization of the spores; yet it was true that, even in comparison to the spores in the first set of letters, the Daschle letter material was exceptionally free of bacterial residue and dispersed easily. The technique for its preparation was sophisticated and pointed to a skilled scientist and a state-sponsored program. When the mailed anthrax was identified as the Ames strain, the

American program, which had years before named and experimented with it still, was further implicated. More tests by the FBI in 2002 showed that the spores had been prepared within the prior two years of their being sent.

At that point, although the FBI appeared to suspect a former Fort Detrick scientist who subsequently worked on intelligence projects, the criminal investigation stalled. The FBI, in developing its profile of a disgruntled American microbiologist as bioterrorist, changed the way that the government looked at this formerly benign category of scientists. Even an American laboratory worker might pose a threat to national security.

Brentwood Postal Employees

The dispersal of anthrax spores put postal workers at serious risk, primarily those where the letters were mechanically processed and coded, at the Hamilton, New Jersey, facility which served Trenton, and at the main Washington, DC, facility called Brentwood after its street address. Of the twenty-two people infected with anthrax, nine were US postal employees. Of the eleven who contracted inhalational anthrax, seven were postal workers. Another victim worked in the State Department mailroom, where the second Washington letter, addressed to Senator Patrick Leahy, had been diverted by mechanical error. Still another victim was an AMI mailroom worker.[18]

In the week after the Daschle letter was opened, evidence of leakage from it and other letters accumulated, but not with the effect of alerting postal workers. On Wednesday, October 17, anthrax spores were found in the mailroom in the adjacent Dirksen Senate Office Building, which served the Hart Building. Senate offices were shut down, as, for cautionary reasons, the House offices had been closed the day before; the Hart building would take five months to decontaminate. In the next two days, there were more warnings. In Florida, a Boca Raton postal worker was diagnosed with cutaneous anthrax. In New Jersey, at the Hamilton facility, nasal swab tests revealed that twelve workers had been exposed to anthrax spores; another worker was diagnosed with cutaneous anthrax. Material from the second set of letters, which seemed to disperse more easily than the first, was the likely source of infection. On Friday, another cutaneous case was diagnosed, which led to the closing of the Hamilton facility and the distribution of antibiotics there.

The US Postal Service, under pressure from Congress to become cost-efficient, had no incentives to close any facility. The 500,000-square-foot Brentwood facility employed nearly two thousand workers and processed much of the federal government's mail. Unaware of the capacity of anthrax spores to disperse and infect at low doses, the CDC officials reckoned that the probability of infection was low. Postal work continued without the distribution of antibiotics or warnings to workers to be alert for symptoms of the disease.

Four of the five inhalational anthrax cases in Washington were Brentwood workers. The first case was that of Leroy Richmond, who was clinically diagnosed on Friday evening, October 19, after being misdiagnosed that afternoon by an HMO physician. This physician, not unlike others during this crisis, made no connection between the patient's workplace, the anthrax crisis, and the patient's flulike symptoms. As the facility that had processed the Daschle letter, Brentwood had been in the news all week. On October 12, within an hour after the Daschle letter was processed through DBCS (Digital Bar Code Sorter) #17, Richmond, who had lent his face mask to another worker, was cleaning the area around the machine when yet another worker, appropriately masked, used air pressure to clean it. Richmond's insistence on a hospital referral brought him to a hospital emergency ward where a physician finally made the connection to Brentwood and alerted the CDC to his case.

On Saturday, October 20, another Brentwood worker was hospitalized and clinically diagnosed with inhalational anthrax. Meanwhile, the CDC waited for laboratory confirmation of Richmond's diagnosis from Atlanta, which was communicated at 7:00 on Sunday morning. That same morning, another floor worker at Brentwood, William Morris, called 911 and died of inhalational anthrax shortly after being admitted to a Washington hospital. Also that morning, a fourth worker, Joseph Curseen, who worked at DBCS #17, collapsed at church. He was diagnosed at a local hospital as suffering from dehydration and sent home. Later he was transported by an emergency ambulance to a hospital where the next day he, too, died of inhalational anthrax.

When the Brentwood postal facility was closed the early afternoon of Sunday, October 21, its distraught employees were aware that three coworkers were hospitalized. Perhaps to maintain calm, the CDC and Washington health officials communicated little sense of urgency in the distribution of antibiotics. A rock concert had blocked access to the nearest hospital, so workers were directed to a downtown building where, from a temporary station, they could get antibiotics and have nasal swabs. Or, officials told them, they could wait and go to the hospital the next day.

The failure to communicate the risks of inhalational anthrax to the postal workers and to describe its symptoms cost lives, as did the lack of clear communication with local physicians to be on the watch for any patients from postal facilities. The CDC and others measured the small numbers of those who got sick or died against the larger numbers who, without antibiotics, might have been casualties. One analyst concluded that the public health response was "rapid and comprehensive, and it may have prevented the further spread of anthrax." The underestimation of danger at Brentwood can also be construed as biomedical example of how organizational thinking can obscure and then increase the likelihood of serious risks.[19]

The random deaths of two more people from inhalational anthrax prolonged national anxiety about postal contamination. Sixty-one-year-old Kathy Nguyen, a hospital worker in New York, died in late October 2001, perhaps from contact with cross-contaminated mail. In late November, ninety-four-year-old Ottalie Lundgren died of inhalational anthrax in rural Connecticut. The post office that served her district proved to be contaminated with anthrax, suggesting that trace amounts on her mail were the likely source of her infection. The spores from the Leahy letter contaminated diplomatic pouches, including one sent to Ekaterinburg (formerly Sverdlovsk). There, to verify the source of the spores, the United States relied on some of the same public health officials who had been involved in containing the 1979 outbreak.

In February 2002, former Brentwood employees, redeployed to other facilities, organized a support group, Brentwood Exposed, to deal with the deaths, illness, stress, and reactions to antibiotics caused by the anthrax letters, for which no federal agency took responsibility. That no perpetrator had been found contributed to the community's loss of trust in law enforcement and government. Their workplace took two years to decontaminate, and many workers were reluctant to return.

Judicial Watch, a Washington legal advocacy group started by a former Justice Department lawyer, took up the case of the Brentwood workers. In December 2002, it acquired a Brentwood manager's diary describing his understanding four days before the facility closed that "hot" anthrax spores had contaminated it. Judicial Watch then filed a class action suit on behalf of the Brentwood workers and pressed for a criminal investigation into the government's handling of the crisis.

The Department of Homeland Security

On November 25, 2003, President Bush signed the Homeland Security Act into law. More than twenty-five agencies were affected, and tens of thousands of government employees would be relocated to the new department. Congress quickly approved Bush's choice for Secretary of Homeland Defense, Tom Ridge.

The stated mission of the new Department of Homeland Security is to "prevent terrorist attacks within the United States, reduce America's vulnerability to terrorism, and to minimize the damage and recover from attacks that do occur."[20] It incorporated the Immigration and Naturalization Service as a way to monitor and protect foreign visitors. It also incorporated FEMA (Federal Emergency Management Agency) and assumed responsibilities unrelated to terrorism—for victims of floods, hurricanes, earthquakes and other providential catastrophes, and for industrial accidents.

Since DHS is charged with mediating intelligence data on threats to the nation and conveying threat levels (red, orange, and yellow, for example) to the nation, the department's activities are often protected by secrecy. Reports of many of its activities are unavailable through Freedom of Information Act requests. Its advisory meetings with contractors are exempt from FACA (Federal Advisory Committee Act) requirements for public disclosure.

The general trend after September 11 was to increase secrecy around federal bureaucracies. The CDC, the nation's public health agency, was for the first time allowed to classify its planning sessions and reports. Counterterrorism activities at the FBI also increased, following a budget increase from $595 million in FY2001 to $1.06 billion in FY2002.[21]

In addition, activity at the Defense Department relative to homeland security expanded. By 2003, the Defense Department had twenty-three different suboffices and groups for WMD disaster response.[22] As part of the military's increased role, in October 2002 US Northern Command (USNORTHCOM) at Peterson Air Force Base in Colorado was established both to coordinate protection of land, maritime, and aerospace approaches to the United States and, in the event of a WMD attack, to providing civil authorities with "technical support" and "assistance to law enforcement; assisting in the restoration of law and order; loaning specialized equipment; [and] assisting in consequence management."[23]

In one effort, the Defense Department also began applying its Advanced Concept Technology Demonstrations (ACTD) for command-and-control military operations to terrorist attack simulations. In December 2002, a complex ACTD exercise simulated twenty related terrorist attacks from New York to Hawaii, with participation by fifty federal agencies, including US marshals, the FBI, the CIA, and the Bureau of Alcohol, Tobacco and Firearms. "The event became the largest real-life, on-the-ground, interagency technical interaction ever conducted for homeland security."[24] The ACTD exercises were later coordinated with DHS as the military's contribution to civil defense, from the federal to the state and local levels.

The DHS maintained the federal commitment to local domestic preparedness, providing funds for emergency personnel, exercises, and equipment and demanding more rational response plans from states. However, due to the economic downturn that began in 2001, the local programs were undercut by layoffs in police, firefighters, and other government workers who routinely safeguard the public. The federal government organization of counter-bioterrorism also began raising ethical and legal concerns about individual citizens' rights during epidemic emergencies.[25]

DHS also continued to evaluate federal organizational response by using simulated attack scenarios. In May 2003, in cooperation with the State Department, it sponsored the TOPOFF 2 exercise, which, like the first one in 2001, was to test how federal agencies performed in emergencies. The scale of the fictive attacks increased geographically from the first TOPOFF, which was city-based, to large northern metropolitan areas that extended into Canada. The TOPOFF budget also increased, from $10 million to $16 million.

The 2003 exercise simulated a radioactive attack in the Seattle metropolitan area and a related pneumonic plague attack affecting greater Chicago. Nineteen federal agencies, the Red Cross, and the government of Canada were involved in the five-day exercise. The fictional perpetrator of the attacks was an international terrorist organization called GLODO. When it was over, aspects of the exercise were classified, an indication that TOPOFF yielded information about national vulnerability that officials deemed should be kept from the nation's enemies and from the American public.

Technological Solutions: The Smallpox Vaccination Campaign

During the second Bush administration, advocates for domestic preparedness and civilian biodefense continued to talk about a revitalization of American public health.[26] Yet public health continued to mean, even more than in the Clinton administration, a technological approach to national defense. In the Bush administration, pharmaceutical protection became the centerpiece of biodefense policy.

On December 13, 2002, convinced of the Dark Winter–type threat of smallpox, President Bush announced his nationwide smallpox inoculation program.[27] Publicity about Iraq's potential biological arsenal, especially in the lead-up to the 2003 invasion, and the threat of bioterrorism had convinced many in the public to participate. The states and the CDC were ready to handle the logistics. In addition, civilian participation was voluntary, which reduced legal liability for those who administered the vaccine and for the government.

As might have been predicted, this smallpox vaccination campaign found it difficult to circumvent the well-known fears of vaccination as a source of bodily pollution[28] and the mistrust engendered when vaccines appear a worse health risk than the forecast epidemic. The swine flu vaccination program during President Gerald Ford's administration was a controversial miscalculation that killed two dozen and sickened hundreds and, because the flu never struck, it caused widespread distrust in government health initiatives.[29] Reliance on intelligence calculations can be a neccessary but problematic basis for predicting epidemics, even for the military. The Pentagon's universal anthrax vaccine program in the 1990s resulted in dishonorable discharge for more than four hundred soldiers who refused to be vaccinated and it caused dozens of National Guard pilots to resign, out of fear of serious, unpredicted side effects that appeared worse than any biological weapons threat.[30]

The first phase of the smallpox campaign was the mandatory vaccination of 500,000 military and government employees who might be deployed to the Middle East or other potentially high-risk areas. Starting January 2003, Phase One also included the voluntary vaccination of 500,000 "front-line" civilian health workers and first responders on specialized "Smallpox Response Teams." In the second phase of the smallpox vaccination program, in March–April 2003, ten million additional healthcare workers and first responders could opt to be vaccinated. In the fall of 2003, in Stage Three the vaccine would become available to the American public at large. (Phases Four and Five were emergency strategies for containment in the event of a smallpox outbreak, which meant quarantine and mass vaccination.)

Unlike the anthrax vaccine, the most available US smallpox vaccine, called Dryvax, had a well-known history.[31] Administered with a bifurcated needle, it took six to eight days to be effective. If given within four days of exposure, the vaccine might significantly reduce the chances of sickness and death.[32] The typical reactions to the smallpox vaccine ranged from soreness at the vaccination site to headaches, swelling of the lymph nodes, and fatigue. Brain swelling (encephalitis) was a known but rare reaction. The death rate estimated for universal smallpox vaccination could theoretically be as high as one percent: if ten million people were vaccinated, some ten thousand might die, unless risk factors were recognized in advance. The vaccine was contraindicated for those with eczema or other skin diseases, for pregnant women, and for those whose immune systems were compromised or who were taking immune-suppressing medication.

Identifying these risk factors posed problems. For example, the CDC estimated that around 300,000 Americans were unknowingly HIV-infected. The smallpox vaccination guidelines suggested HIV testing but did not insist on this precaution. In addition, after the shot, the vaccine site could shed virus cells and cause illness (called "contact vaccinia") that could be dangerous to others.

Throughout 2003, just a few reported adverse reactions to the smallpox vaccine caused a drop in public participation. Unexpected heart symptoms (cardiac adverse events) were particularly alarming.[33] Among the 250,000 soldiers vaccinated for the first time by March 31, fourteen (ranging in ages from twenty-one to thirty-three) suffered heart problems, either myocarditis or pericarditis or both. The 100,000 other soldiers being revaccinated reported no such problems with inflammation. Overall, the military was positive about the smallpox vaccine program.

By late March 2003, just under thirty thousand civilians had been vaccinated nationally, a small fraction of those expected to cooperate. Three of these volunteers had heart attacks, two of them fatal, and seven others suffered other heart-related problems. The CDC characterized the three first responders (ages fifty-five to sixty-four) who suffered heart attacks as already having "clearly defined risk factors," such as high cholesterol levels, cigarette smoking, and a previous history of heart trouble. The same appeared true for a fifty-five-year-old member of the National Guard who died of a heart attack five days after being vaccinated. The distinction between the vaccine's causing heart failure and its contributing to heart failure was lost on many. The threat of terrorists attacking with smallpox aerosol seemed more remote than these reported illnesses and deaths, especially as the Iraq war was declared over in May and the threat of bioterrorism faded from the news. Fifteen states immediately halted their vaccination programs. Many hospitals independently withdrew participation.

The federal government stayed committed to the program, although the intelligence data supporting this commitment, always vague, remained unspecific. In July 2003, the CDC was given $100 million to dispense to states to improve participation rates. The Johns Hopkins physicians who had organized the Dark Winter scenario remained optimistic about the smallpox campaign as a way to reduce the risks of bioterrorism.[34]

Science for Biodefense

The Bioterrorism Act of 2002 was the comprehensive Congressional response to defend "public health security" and the strongest vote yet for technological solutions to the perceived threat.[35] This legislation mandated increased surveillance of drinking water supplies and food and required secret stockpiles of drugs, vaccines (especially smallpox vaccine), and other emergency medical supplies. It defined the importance of "accelerated countermeasure research and development," that is, research on pathogens of potential use in a bioterrorist attack, with the Secretary of Health and Human Services tasked to fund research on them, in collaboration with the Department of Defense and the Joint Genome Institute of the Department of Energy. The Department of Veterans Affairs, with its large hospital system, was suggested as a biomedical research and development resource and was allocated $33 million for this purpose. The bill affirmed that "accelerating the crucial work done at university centers and laboratories will contribute significantly to the United States capacity to defend against any biological threat or attack."

The distribution of funds in the 2004 Department of Homeland Security Budget ($29.7 billion total) signified the importance attached to technological innovation as a way to counter the threat of asymmetric WMD attacks on civilians. Despite the unsuccessful smallpox vaccination program, the largest single budget item was for Bioshield, the Bush administration's program for the development of better vaccines and drugs by pharmaceutical companies. Bioshield, a biomedical equivalent of Reagan's Star Wars program, promised universal protection from biological weapons but faced the uncertainty of the threat, the technology, and organization for national vaccinations or other campaigns. Pharmaceutical companies would receive $5.6 billion over ten years, with $890 million allocated for 2004. The budget allocated $455 million for technologies to counter biological, radiological, chemical, nuclear, and high explosives attacks on civilians. The budget included $40 million for biosensors in urban areas; $50 million was earmarked for the Metropolitan Medical Response System to enhance disease reporting.

Specifically to combat bioterrorism, the Bush administration directed important new funding to biomedical science. At the National Institutes of Health, the National Institute of Allergy and Infectious Diseases (NIAID), the largest research organization of its kind in the world and perhaps the most diverse and advanced in its projects, received $1.7 billion (of its total $4 billion budget) in new money to jumpstart the invention of new vaccine, antibiotics, and technologies for early diagnosis. Through Homeland Security, $70 million was also earmarked for eight to ten Regional Centers of Excellence located in major research universities and institutes. The number of Level 4 containment laboratories for research on dangerous pathogens was also slated to increase.

Although NIAID would focus on protection against the best known select agents (smallpox, anthrax, tularemia, plague, botulinum toxin, and hemorrhagic fever viruses), Anthony Fauci, NIAID's director since 1984, claimed that important discoveries about infectious diseases in general could be made while investigating pathogens that posed little or no current public health threat. For example, the study of smallpox or Ebola might yield an antiviral drug that worked against other viruses, or research on the anthrax toxins might yield an antitoxin with alternate uses. Fauci compared the transition to counter-bioterrorism research to the development of AIDS research in the 1980s, when the public and young scientists had to be educated about its importance. He concluded that "there needs to be an accelerated and sustained effort to draw more competent scientists/researchers into the field of research on counter-bioterrorism."[36]

The integration of basic science into the biodefense project meant that some university and medical center scientists would confront national security restraints not experienced since World War II or during the US offensive program before it ended with the 1969 Nixon decision. The emphasis would be on regulating pathogens, facilities, and scientists in the public realm. Mild restrictions were already in place in the 1990s; for example, the 1996 Anti-Terrorism and Effective Death Penalty Act required official reporting to CDC if listed pathogens were transferred from one laboratory to another. At that time, the presumption was that all biomedical scientists worked against disease. The Bioterrorism Act of 2002 required authorities at a research or medical facility to report work with any of the forty-two CDC select agents. The government would keep this bank of pathogen information secret, except in public health emergencies (presumably accidents, bioterrorist attacks, or other unusual outbreaks) or for congressional committees.

Authorities also had to submit the names of employees with a legitimate need for access to the pathogens to the secretary of HHS and the office of the US Attorney General. The names would then be run through criminal, immigration, national security, and other electronic databases to check for possible "restricted person" status. "Restricted persons" had been defined in the US Patriot Act of October 2001 (PL 107–56) as felons, fugitives from justice, those dishonorably discharged from the armed services, the mentally defective, illegal aliens, and foreign nationals from terrorism-sponsoring nations.[37] By the new law, the attorney general could identify any individual as suspect who had been named in a warrant for violent criminal or terrorist activity or was suspected of spying for the military or intelligence operations of a foreign nation. These regulations applied to open nongovernmental laboratories, with the presumption that the government was monitoring its own classified research on dangerous pathogens.

To defend the nation against biological weapons—wherever they were and whoever had them—responsible biologists were asked to give up some measure of openness and free communication or choose other research paths. No matter what research they chose, some new technique or discovery might if published lend itself to an enemy's advantage. Quite suddenly, the entire enterprise of biology became potentially suspect. As early as November 2001 an editorial in the science journal *Nature* asked whether, with bioterrorism looming, the biological sciences had reached the end of innocence.[38]

Between September 2001 and the end of 2003, the United States government developed an aggressive, nationalistic posture toward the potential threat of biological weapons. It invaded Iraq with the goal, among others, of eliminating its biological weapons threat; it created a new federal department for civil defense, with a considerable emphasis on biodefense; and it sought to conform an important segment of American biology to biodefense via pharmaceuticals and other technologies.

These post–9/11 US initiatives will have organizational repercussions for many years. In the next chapter, we review their possible consequences and consider other restraints available to protect civilians from biological weapons.

CHAPTER 10

Biological Weapons

Restraints Against Proliferation

Biology today is as susceptible to hostile exploitation as were chemistry in World War I and physics in World War II. The formidable power of international commerce is behind this basic science, moving it toward innovations that, along with marketable medical value, might also be turned to destructive ends. If exploited by states, the science and technology of biological weapons could pose one of the most serious problems humanity has ever faced. A new generation of biological weapons, if pursued with vigor, could make them technologically competitive, especially for human control and domination. Unless the power of biotechnology is politically restrained, it could introduce scientific methods that would change the way war is waged and increase the means for victimizing civilians.

The question at present is whether sufficient national and international restraints against this danger are in place, especially when scientific knowledge itself is at issue. In the past, various, at times serendipitous, combinations of legal norms, public oversight, technical obstacles, and political leadership prevented the use of biological weapons. Overall, though, the world has been lucky, in that influential political actors took action at critical junctures and that the general historical trend of the last century has been toward transparency and open government.

History shows that the problem of biological weapons proliferation is too complex to be solved by any single restraint.[1] A reasoned assessment of threats and measures is the first step to resolving the problem of proliferation. This problem should be understood as potentially more serious now than in the past, in great measure because human malice coupled with human ingenuity is a constant in world history. History also tells us that without a long-term commitment to nonproliferation, we are gambling with the future.[2]

Trust and Mistrust

The United States' definition of its interests sets a standard to which the rest of the world cannot help but react. Even before September 11 and the 2001 anthrax letter attacks, the United States was in retreat from new international initiatives to strengthen the Biological Weapons Convention (BWC), while it also reinforced its homeland security policies. The Bush administration reinforced American unilateralism and a confrontational approach to international relations that, as in the Reagan era, depended on conspicuous military might. After September 11, "waging war on two fronts" was President Bush's apt description for America's militancy abroad and civil-defense orientation at home.[3]

At the core of the US government's rejection of a strengthened BWC was the belief that the United States was exceptionally trustworthy and could therefore interpret and implement legal norms as it chose. Close allies of the United States, especially the United Kingdom, were trustworthy as well. In distinct contrast, suspect nations such as Iraq, North Korea, Libya, Syria, and Iran posed the most serious threat because they were too closed and perfidious to abide by law. Other states, such as China, India, Pakistan, and later Russia, were too well armed and large to ignore but still not fully trusted.

This perception of a global division between "haves" and "have-nots" is often shared among advanced industrial nations and has influenced treaty negotiations on weapons of mass destruction. From this viewpoint, the world appears divided into "responsible" Western states that can be trusted with nuclear weapons and "irresponsible" non-Western states that cannot be trusted with the weapons they have or that must be prevented from acquiring any. In this view, arms control is but one element of a web of restraints imposed by the North on the South, encompassing export controls, strong biological and chemical defenses, and a "determined and effective" military response.[4]

A competing political vision assumes inseparable common interests among all the world's states in reducing the threat of weapons of mass destruction and of war and violence in general. It holds that the benefits of biomedical technology should be shared between wealthy nations and those in economic need. This perspective promotes the international mobilization of citizens, including scientists and physicians, to promote the interdiction of biological weapons. Far from being antagonistic to American or Western values, this approach is based on openness and democratic participation, extended to a global context. It encourages nongovernmental and grassroots organizations and the idea of civil society as part of the long-term international solution to present dangers.[5]

The dichotomy between the two approaches is less sharp than it may appear. Despite the forces of globalization, the importance of sovereign state governments and state law in reducing weapons proliferation endures. The initiatives of advanced industrial states can be vitally constructive, provided there are options for broader cooperation. International measures based on trust are likewise essential for persuading a range of governments that openness through, for example, mutual onsite inspections and scientific partnership, is preferable to continued secrecy.

The BWC and US Military Secrecy

At present 151 nations are parties to the Biological Weapons Convention, and 132 nations are parties to the Geneva Protocol, a strong testimony to the international norm. The BWC's lack of verification and compliance provisions is frequently referred to as its greatest shortcoming in reducing the threat of proliferation. What, then, is the solution?

US government objections to a protocol that would strengthen the treaty have been based on what it has argued is its exceptional need for military secrecy and the proprietary needs of its pharmaceutical industries. This double argument raises the question of whether, regarding this category of weapon, the safety of one nation is strictly divisible from all others, even, for example, from that of European allies like the United Kingdom, which have supported treaty compliance and transparency.

The US defensive program is the largest in the world, a fact that US representatives have often underscored. But how large and, more important, how secret should it be to serve national interests—which are fundamentally public interests—or those of its allies? The 1991 BWC Confidence-Building Measures, to which the US government agreed, require declarations of legitimate defensive projects and locales, leaving the substance of the classified work undisturbed. These declarations have compliance, not total transparency, as a goal. Further, they are intended as a means of distinguishing between legitimate programs and the illegitimate ventures that are more likely to be hidden and therefore undeclared.[6] The United States should have no need for biological defense programs whose locations and general nature must be kept secret. But how much secrecy is too much?

If, as the covert CIA and Department of Defense projects suggest, the United States has a stake in undeclared projects at undeclared sites, it may be engaged in activities that increase the risks of proliferation and therefore danger to the public. With the world's largest military, the United States could set broad and dangerous standards for biological weapons research. Escalation within a defensive program, for example, could advance the laboratory and delivery technology for biological weapons. Small-scale, arbitrary projects, such as those leaked to the press in 2001, could pave the way for new agents or for attack simulations on a large, elaborate scale, to second-guess what a suspected enemy might develop. In time, other states or organizations would gain access to the same or similar classified technology, whether for a new agent, drug, bomb, or missile.

The US presumption that the threat of biological weapons is foreign is generally supportable, but several domestic crimes, including the 2001 anthrax letters, point to risks within the United States, perhaps from its own programs. Intelligence and defense officials may want to explore what an adversary might do, but secret knowledge and materials generated by their own biodefense projects could be put to hostile use by the increasing numbers of Americans with access and special skills.

Transparency in the name of public safety need not require front-page headlines, but it does require oversight and accountability. Ample US military funding without civil review fueled biological weapons proliferation in the past, and it could pose a danger again.

Legal Restraints and the Pharmaceutical Industry

The second US objection to strengthening the BWC concerns the protection of the proprietary rights of pharmaceutical companies. During negotiations on the BWC protocol, the lobby for the American pharmaceutical and biotechnology industry, Pharmaceutical Research and Manufacturers of America (PhRMA), was vocal in its objections to onsite inspections by international teams as an economic threat.[7]

The cooperation of American pharmaceutical companies is vital to an effective BWC. At present their position resembles that of the American chemical industry in 1925 when the Geneva Protocol ratification was debated in the Senate. Competition then was strong among American companies and between American industry and European chemical manufacturers, while the non-Western world of vast colonies and emerging states lagged far behind in industrial capacity and consumption.

During the 1990s, pharmaceutical and biotechnology industries were beginning to reap unprecedented rewards from years of basic research, much of it funded by the federal government. Earlier public fears about experiments with recombinant DNA had been resolved by the creation of institutional and federal government oversight committees.[8] The exploitation of revolutionary technological advances, for example, gene-splicing and genome sequence databanks, promised to be highly profitable. In this competitive atmosphere, protecting patents and confidential business information was primary.

Predictably, the probusiness Bush administration defended the industry's right to secrecy. The pharmaceutical industry, though, is different from others in that its products directly promote health and save lives and therefore have a value that is universally subsidized by advanced industrialized nations for their citizens. That said, these corporations have responsibilities to stockholders and they protect their turf. For example, nearly every major pharmaceutical company joined in a 2001 suit against the government of South Africa's intent to import cheap generic AIDS medications, which the industry feared would set a global precedent.[9] The bad press that followed this suit, which was subsequently dropped, motivated a number of the giant corporations to work with the World Health Organization to provide inexpensive, subsidized HIV medications for needy nations.

Nations differ greatly in the accountability they demand of their pharmaceutical industry and the controls they impose on them. Western European states, perhaps uniquely in the world, have required openness and accountability in the name of the public good. Perhaps as a consequence, Western European pharmaceutical companies were more accepting of protocol declaration and inspection proposals than PhRMA. In the United Kingdom, a strong supporter of the protocol, the political context for emerging biotechnologies allowed the public and environmentalists to review laboratory and manufacturing processes. This emphasis on social participation grew out of the government's assumption of responsibility for health and safety risks during World War II, which evolved into the United Kingdom's socialized medicine based on local citizen councils. In Germany, pharmaceutical accountability to the public has been accorded even higher priority, for the general environmental risks that manufacturers might cause.[10]

In contrast, the American approach to biotechnology has been to shield the manufacturing processes from oversight and to emphasize the product, which is judged by federal regulators and in the marketplace, where individual consumers might seek legal compensation if they have grievances.

This American emphasis on product helps explain US industry's resistance to the BWC protocol. Yet, throughout the years of Ad Hoc Group discussions, the emphasis was on inspection with regard to large-scale production, which left ample room for the protection of commercial secrets during sensitive research and development leading to patents. Even this end-of-process verification, though, proved unsatisfactory to the United States.[11]

With the US Bioshield program and increased funding for the National Institute of Allergy and Infectious Diseases (NIAID), American pharmaceutical companies now have a stake in civil defense that bears watching. The companies involved in biodefense are in a more precarious position than their officers may realize. They stand to gain only if there is a technical invention from basic research, for instance, a generally effective antiviral drug or other discovery beyond traditional defenses against the relatively arcane biological weapons agents. If pharmaceutical companies are drawn into national vaccine or drug distribution campaigns, they could emerge as heroes or, if there is a false alarm or exaggerated risk, they might be seen as purveyors of sickness and death. Even before that, a spate of laboratory accidents, a product that causes injury in tests, a scientist who perpetrates a biocrime, extensive monkey and other animal research—any of these could harm the industry's image. In the long run, corporate leaders in biotechnology may eventually want to assume a larger role in preventing biological weapons proliferation, as the US chemical industry did after Vietnam and the use of chemicals in that war.[12]

Preoccupied with defending American pharmaceuticals, the United States failed to consider how other nations—for example, China, Russia, India, and Brazil—with state-owned and state-protected biotechnology companies might be susceptible to military exploitation. The potential for "niche states" developing second, third, and further generations of biological weapons would be greater in the absence of verification and compliance measures. Iraq and South Africa are examples of program proliferation in an era of political indifference to their activities. Is such indifference supportable now?

The Concept of Legal Completion

The United Kingdom, vigorously supporting the BWC protocol and legal restraints, argued that without pressure on all states to agree to verification measures (declarations and inspections) the opportunities for terrorists would increase worldwide: "Improving states' control of biological and toxin agents is a necessary component of international co-operation to ensure that they do not fall into the hands of terrorists."[13] States have legal means to restrain potential proliferation, if they are called upon to use them. British scholar and BWC analyst Nicholas Sims has identified five legal treaty requirements approved by its review conferences but with which not all states that are party to the BWC have complied.[14]

The first of these agreements concerns Article IV, which calls on states to adopt national legislation to implement their obligations to the treaty. One such obligation is agreed to be the enactment of national penal legislation regarding activities that are prohibited by the Convention.

In 1991, states agreed to submit Form E to report annually the status of their compliance. National legal implementation has been uneven and slow (the United States passed its legislation in 1989) and, since it constitutes disregard of a BWC legal requirement, represents a threat to treaty compliance. A recent survey by VERTIC (Verification Research, Training and Information Centre) in London shows that 47 percent of the 146 states parties have enacted an enabling law, with another 7 percent in the process of acquiring one.[15] Africa was least in conformity, with only 16 percent of its nations having such laws, whereas in Europe, 73 percent of its states had passed the required legislation. Fifty-six states (37 percent of the world total) had no information to provide to VERTIC, nearly all of them nations in troubled world areas. Pakistan, Sudan, and North Korea were on this list, along with Saudi Arabia, Qatar, and Indonesia.

Second, a requirement to share information on national measures is related to national implementing legislation and also required by Article IV. In communicating the means by which domestic legislation follows the BWC ban, a state demonstrates that it has translated the ban effectively and stringently, and it adds to a worldwide network of standard legal restraints.

Third, regarding Article VIII of the BWC, the first review conference declaration urged all states parties that had not already done so to become party to the Geneva Protocol. Thus far, twenty-five states parties to the BWC have not ratified or acceded to the Geneva Protocol. Again, most of them are in troubled regions of the world, the Middle East, Africa, Eastern Europe, and South America. In Sims's words, refusing to accept the Geneva Protocol sends "a message of equivocation" that weakens the BWC.

Fourth, the reservations to the Geneva Protocol that maintained an option for retaliation in kind—which permitted the old French and British programs—should have been withdrawn as states became party to the BWC. The terms of Article I are absolute: development, production, stockpiling, acquisition, and retention of biological and toxin weapons are forbidden "with the aim to exclude completely and forever the possibility of their use." About twenty state parties, including China, India, and North Korea, have retained reservations that might be understood as disdainful of the BWC.

Finally, Article IX of the BWC calls for states to support negotiations to achieve a total ban on chemical weapons. Today, this is taken as an obligation to become party to the Chemical Weapons Convention (CWC), which is the result of those negotiations. Nevertheless, some dozen states party to the BWC have not joined the CWC.

International Criminalization of Chemical and Biological Weapons

The legal gap left by the various states' lack of national legislation in compliance with Article IV is compounded by the lack of international law to criminalize biological and chemical weapons. Closing this loophole would make it difficult for a perpetrator to escape jurisdiction in a state's domestic court should that individual be found in a state that supports the law.

Beginning in 1996, a group of scientists and lawyers began formulating a new legal convention that would combine the prohibitions of the BWC and the CWC to make it a crime under international law for any person to develop, produce, acquire, stockpile, retain, or use chemical or biological weapons; to assist, encourage, induce, order, or direct anyone to engage in any such activities; or to threaten or engage in any preparations to use such weapons.[16]

Seven international treaties already in force offer instructive models. Each provides the sort of expanded jurisdiction that could be incorporated into a new treaty for chemical and biological weapons crimes, so that it makes no difference in what country the crime was committed. Existing treaties such as the 1970 Airline Hijacking Convention, the 1979 Hostage-Taking Convention, and the 1984 Convention Against Torture apply in any participating state in which the accused offender is found, even if the state has no links of territory with the offense or of nationality with the offender or with the victim or victims. Thus, in 1998, the membership of the United Kingdom in the Torture Convention made it possible for former President Augusto Pinochet of Chile to be detained in London for extradition to Spain.

Rather than identifying specific biological agents, the proposed criminalization treaty follows both the BWC and the CWC in their emphasis on hostile intent, that is, the purpose of the activity. In this way, technological change and the development of new agents and means of delivery are anticipated and the commercial use of many substances is protected. Provisions for dispute resolution and due process are part of the treaty proposal, as they are for similar treaties pertaining to crimes against humanity.

The prospects for this type of criminal treaty reflect the world climate for international justice. While the US administration is opposed to the establishment of the International Criminal Court, many other nations, including its close allies, have joined it. The prosecution of war crimes has a well established, if controversial, history from World War II. This history now includes tribunals established by the UN Security Council for prosecuting individuals for the mass murders that took place in Bosnia and Rwanda during the 1990s and in the former Yugoslavia in the 1980s. In 1998, the Rome Statute of the International Criminal Court introduced criminal penalties for the use in war of chemical weapons, using the phrase from the Geneva Protocol that bans "asphyxiating, poisonous or other gases, and all analogous liquids, materials or devices." It also points to state-level responsibility for infractions, defining such actions as those "committed as part of a plan or policy or as part of a large-scale commission of such crimes."[17]

The fortification of legal restraints against biological weapons proliferation is bound to be a slow and difficult process of reinforcing norms in widely divergent contexts. Since the threat of biological weapons is global, then international legal initiatives are essential in order to protect US civilians and civilians everywhere, the primary targets of bioterrorism. Like all laws, they can succeed only if backed by legitimate authority and reinforced by other restraints.

Trust and International Control of Dangerous Pathogens

The control of dangerous biological pathogens to restrain proliferation has been approached in a variety of ways, often with an emphasis on trade restrictions. In the 1980s, in response to Iraq's use of chemical weapons against Iran, an organization of advanced industrial nations, now called the Australia Group (AG), assembled in Paris to agree on export controls on chemical weapons agents and their precursors and technologies. The AG later expanded its controls to include biological agents and equipment.

Through licensing measures on the export of certain chemicals, biological agents, and dual use equipment, the Australia Group's objective is to keep these items from contributing to the proliferation of chemical and biological weapons. By 2000, fifty-one pathogenic microbes and nineteen toxins were on the core list of agents subject to the group's export controls.

The AG clearly divides the world into haves and have-nots. None of its thirty-four members falls in the category of poor developing nation; some excluded nations see threats to their economies in the Australia Group restrictions on dual-use technologies.[18]

The United States has resorted to trade sanctions to control specific lesser states. In 1991, before its relations with Iraq deteriorated, the United States passed the Chemical and Biological Weapons Control Act, which promised that it would cease trade with any state that exported agents or equipment contributing to proliferation. This legislation specifically banned such exports to Iran, Libya, Syria, North Korea, and Cuba. In 1992 the US Congress passed the Iran-Iraq Arms Nonproliferation Act with sanctions against trade that could lead to WMD capability. In 2000, the Iran Nonproliferation Act specifically targeted Iran and sanctioned foreign persons trading in WMD technology, which included the Australia Group lists of agents and those of the 1993 Chemical Weapons Convention.

In 2002, the Bush administration approached controlling WMD technologies with a similar model, this time to encourage intelligence sharing among the major powers as well as the interdiction of shipments of WMD materials and equipment. On May 31, 2003, President Bush announced the Proliferation Security Initiative (PSI), a cooperative venture with nine European allies, along with Australia and Japan, to interdict illegal traffic relating to WMD at sea, in the air, or on land, and to streamline a rapid exchange of relevant intelligence.[19] Within a few months, a series of simulations of interdiction raids began, by Australia in the Coral Sea, by Spain in the Western Mediterranean, and by France off its Mediterranean coast, and a tabletop air exercise by the United Kingdom. Described as an "activity" rather than an organization, PSI was intended to create "a web of counterproliferation partnerships through which proliferators will have difficulty carrying out their trade in WMD and missile-related technology."[20]

The kind of trade control represented by the AG and the PSI may be less effective in relation to biological weapons agents and equipment than it is for either chemical or nuclear arms. Microbial pathogens, of which there are many in nature, can be transferred in minute amounts and later made to proliferate in quantity. The equipment needed to grow and process biological agents, as well as growth media, would be generic. Those in charge of raids would likely have preconceptions about violators based on their nationalities but might have difficulty discovering proof of culpability.

Other proposals regarding the control of pathogens have avoided the thorny issue of trust among nations by focusing on scientific authority over research with dangerous pathogens. In one plan, tiers of scientific peer review could be built on institutional and national mechanisms already in place to control recombinant DNA research and research on laboratory animals and human subjects. In the 1970s, the US National Institutes of Health established the Recombinant DNA Advisory Committee (RAC) to review potentially hazardous research, with local research institutions gradually assuming control. Above such a national structure, an international organization, analogous to the WHO Advisory Committee on Variola Virus, which oversees US and Russian smallpox research, could monitor research.[21] This plan and others for international monitoring of research are more directed at unintended biological hazards than at secret deliberate activities; but, by increasing transparency and creating international networks, they might also act to restrain clandestine efforts. The effective application of such monitoring to secret but presumably legitimate biodefense work would have to overcome US opposition to any kind of international oversight of its activities in this area.

Homeland Security and National Science: Public Trust

During the BWC Protocol negotiations the Pentagon claimed that its role in defending the civilian population against bioterrorism justified the US requirement for secrecy.[22] As the Department of Homeland Security (DHS) became established, it incorporated most Department of Defense programs related to domestic preparedness, as well as those from the Department of Energy and Department of Agriculture and elsewhere in the federal government.

The DHS is a new and experimental venture whose relations with other departments and agencies are still evolving. As its organizational culture takes shape, the greatest challenge to DHS is the cultivation of public trust, which is necessary to its mission to mobilize communities on their own behalf. Trust in the department depends on its credibility in representing risks. If the DHS is secretive and adamantly closed to public inquiry, its accountability to civilian welfare can become suspect.

Regarding accountability, DHS has inherited a domestic preparedness program from the 1990s that needs reevaluation and change. One problem is the way in which domestic preparedness restricts the public to highly reactive roles, as "responders" or as victims in fictive WMD attacks, instead of encouraging the initiative or feedback that serves best in responding to disaster events.[23] In addition, due to layoffs in local police and firefighters, the ranks of first responders have thinned. A state might have substantial federal funds for bioterrorism response training and many less police on the street. This change, even as it lowers routine security in communities, tips the safeguarding of the public to emergency federal management, which was never intended to make up the difference. What autonomy do states and cities have, then, in deciding what domestic preparedness and homeland security mean?

Another problem looming for the DHS could be a bureaucratic loss of perspective on actual dangers, as crisis management becomes normalized. Political scientist Thomas Schelling, commenting on the 1941 Japanese attack on Pearl Harbor, remarked that government agencies tend to foster "a poverty of expectations—a routine obsession with a few dangers that may be familiar rather than likely."[24] Since the September 11 attacks and the anthrax letters, federal officials have learned to communicate with each other despite agency differences in goals and methods. The creation of the DHS has accelerated that communication, with an increase in shared professional experience across the board. The pitfall may be that dedicated government officials concentrate on daily responsibilities in a rational environment, often ignoring unusual ideas or new frameworks for problem solving or perceiving risk.

Inside and outside DHS, federal officials now involved in homeland security must also reckon with intelligence data, who has access to it and who does not, and its interpretation. Secrets further complicate planning for local responses to unusual disasters and outbreaks of disease. The sharing of information up and down official lines of communication and across agencies is crucial in an emergency and should extend to the public. What, for example, does a FEMA (Federal Emergency Management Agency) official in DHS know that a local physician responding to an unusual outbreak does not? Or, to phrase the question differently, is there any government information about an unusual disease occurrence that should be withheld from citizens who must make decisions to protect themselves and their families? The answer is likely none at all. The 1979 Sverdlovsk outbreak and the postal deaths from the 2001 anthrax letters both demonstrated the increased risks that secrecy and misinformation impose on the public. Future outbreaks may again show that information barriers cost lives while accurate communication reduces risks.

Although largely ignored in current policy, the importance of basic, accessible, and good-quality health services, a combination of public health programs and clinical care, would be the best protection against an unusual deadly outbreak. This was Theodor Rosebury's argument in 1942, and it has merit today. Epidemics are opportunistic, targeting the more vulnerable populations. The better the general health levels in society, so that disparities are diminished, the less destructive the impact of a large disease outbreak would be.

There are other benefits to basic, accessible health care. In an emergency, people should know in advance where to go for protective medical interventions and how to get there quickly or make contact. For that, it helps greatly if they are already getting primary care from a local clinic, HMO, or hospital. They should also trust the medical personnel there who, in an emergency, would be in charge.

Similarly, informed participation at the local level is crucial in disaster situations and especially so with serious epidemics, which, to be defeated, demand public comprehension of how diseases are transmitted and what to do over the course of days or weeks. Before any unusual outbreak, the public should be educated about its central role in routinely preventing and containing infectious diseases and its members should understand disease risks and symptoms. No civil defense mobilization is required for this level of protection; good multilingual public health programs in community centers, clinics, and schools, via the mass media or electronically, would suffice. Without informed public cooperation, the American technological investment—in new laboratories and biological research, in biosensors, drug and vaccine stockpiles, and electronic tracking of disease—is nearly useless.

The strictly national agenda of DHS to control bioterrorism is sure to be challenged. The risks of a serious US epidemic from any source would be difficult to separate from the risks and repercussions felt in other nations. It is difficult, too, to predict which other nations, wealthy or poor or in between, will be affected by epidemics. For years, infectious-disease experts have warned of the international dangers of emerging infectious diseases, of which AIDS is the foremost example and the 2003 SARS (Severe Acute Respiratory Syndrome) outbreak among the most recent. As a recent Institute of Medicine report documents, the shared risks of serious outbreaks are more widespread than once thought.[25] In an age of fast and frequent travel among nations, epizootics and epidemics travel across national boundaries, as witness the spread of West Nile virus across America, the annual influenza season, and the 2003 SARS epidemic. As with large epidemics of the past, today's preventable epidemics of AIDS, malaria, tuberculosis, cholera, and other diseases, which kill millions every year, are part of the chronic cycles of poverty and violence that generate both despair and terrorism.[26] This degradation of human life parallels the devaluation of human life on which total war and strategic weapons were based. Federal officials newly engaged in homeland security and biodefense may find themselves also contemplating the wider social reality of epidemics.[27]

US Biologists and Biodefense

Thus far, in confronting bioterrorism, federal policy has emphasized ingenious technological solutions. For example, DHS has instituted the Homeland Security Advanced Research Projects Agency (HSARPA), modeled on a similar agency at the Department of Defense (DARPA) that funds extramural research on innovative defense technologies.

The greater part of the technology investment is in basic biomedical sciences, also for "magic bullets" to counter biological agents. In 2004, half the total federal budget for homeland security research and development (around $2 billion) was allocated to the National Institute for Allergy and Infectious Diseases for its own laboratories and for extramural projects to create pharmaceutical products.[28] In this way, the civil defense goal to protect all Americans and a probusiness policy came together. The assumption on which this technology-driven policy was based is that, with only general reference to an actual bioterrorist threat, scientific research can defeat each established biological agent or their permutations or perhaps achieve defenses applicable to entire categories of diseases or toxins.

The US government's quest to incorporate biology and related sciences into its civil defense agenda heralds a significant change for scientists in these areas. As early as March 2002, the Defense Department was circulating proposed restrictions on the aims and conduct of scientific research, as a means of controlling the transfer of biological weapons technology. This reaction, fitting for other types of weapons, could barely begin to address the large, unwieldy field of biotechnology, which is as international as it is American. At one point, the FBI estimated to Congress that 22,000 US laboratories might harbor dangerous pathogens in old, leftover stocks from research on outbreaks and epizootics.[29] In theory, any scientist trained in biological laboratory techniques might undermine national security if determined to do so. The federal requirements for laboratory and personnel registration written into the 2002 Bioterrorism Act were a comprehensive step toward controlling the potential threat. The same legislation implemented the funding of biologists to invent solutions to the bioterrorism threat, which would bring more scientific research under government supervision and align it with the larger military defense and probusiness framework for national security.

Few US scientists in infectious disease research have much experience with secret government activities or with restrictions on the choice or substance of their work or the publication of its results. Unlike nuclear physicists and engineers, biologists have intended exclusively beneficent ends to their research, ideally with universal application.

Through peer-review processes, researchers in biology have been in charge of the distribution of generous federal funding since the end of World War II, with few strings attached. At times coercive federal politics did intrude. In the 1950s, for example, Elvin Kabat, coauthor of the Rosebury-Kabat report, was blacklisted after a research colleague alleged that he was a Communist. As a result, despite his contribution to the war effort and consulting for the program that lasted until 1947, he lost federal research funds for several years. Scientific freedom from government intrusion (although not from government regulation) has been much more characteristic, as in the example of President Reagan's National Security Decision Directive 189, which affirmed that the products of fundamental research should remain unrestricted unless they were part of classified projects.[30] This decision guaranteed scientists open review and acknowledgment of their research, while increasing the odds that government funding would go to top-level, creative performers.

For biologists and other scientists engaged in biodefense research, decisions about the goals of their research are subject to the national security agenda, which gives priority to the interests of the United States over all other considerations. The protection of American civilians, the stated biodefense goal, limits beneficence to one population and presumably withholds it from enemies, however broadly or narrowly they are defined. People in North Korea and Iran may be on that list. People in mainland China or Pakistan or other countries could be added, as political will determines.

After the Bioterrorism Act of 2002 and the infusion of new federal money into NIAID, the possible imposition of national security restrictions started a controversy involving DHS, the Department of Defense, the NIH, and scientific organizations, most notably the American Society for Microbiology (ASM). With billions of dollars per year newly directed to research on anthrax, smallpox, and other select agents, biologists have been asked to look at their research for the harm it might promote and the possible necessity to categorize it as classified or as "sensitive but unclassified." If the parameters for evaluation were broadly set, national security threats might be read into many research articles, even those not intentionally related to biological weapons. A new laboratory technique, a new mechanical device, or an unforeseen discovery about a pathogen or drug could all be withheld from publication, to the detriment of the researchers, their institutions, and science in general. Predictably, the possible top-down federal censorship of research publications became a contentious issue. Should scientists instead self-censor their work, or should total openness prevail?

Contributing to this controversy were three original research articles, one regarding an engineered mousepox virus of enhanced virulence in laboratory animals, another on the production of an infectious poliovirus from laboratory reagents, and a third on enhancing the virulence of the vaccinia virus.[31] To some minds, such research could inspire terrorists and should not have been published. Among scientists, the opposing argument was that having the information made public increased the chances of countering potentially dangerous findings with more research and that this problem solving would happen faster, since more scientists would be involved. This debate took place with little questioning of why terrorists, who have had access to biological agents and small delivery technologies for decades, have rarely resorted to them.

The ASM, which publishes eleven scientific journals, chose to monitor scientific research reports and instituted procedures for its reviewers to address potential national security risks in articles, primarily those on select agents. In addition, in February 2003, the editors of major science journals, including *Nature* and *Science*, published a joint statement, "Scientific Publication and Security," to affirm their responsibility for monitoring submitted articles. They acknowledged that they might at times have to forbid publication of an article because of the potential "harm" it might cause, although the criteria for such censorship remained unformulated.[32]

Clamping down on potentially dangerous scientific knowledge promises to be difficult, if not impossible. Ten thousand life science journals are published worldwide, the annual article submissions have recently reached nearly a half million per year, and as many as six thousand such journals are now available electronically, with assistance from the National Institutes of Health.[33] American citizens are not necessarily in charge of all major journals or professional organizations, nor are they the majority of those who submit research articles. Intense and pervasive censorship would surely change this flow of information, with unpredictable consequences.

If research in biology might be used by terrorists or states to harm the public, what control can scientists collectively exert? A 2003 National Research Council (NRC) special committee report took the position that scientists should control judgments of what research constitutes a national security threat.[34] Its major organizational proposal was to make the RAC the basic model for a layered system of oversight, to define "Experiments of Concern" that, like the development of a vaccine-resistant strain of pathogen, should be peer reviewed before being undertaken or published in full detail.[35]

Individual scientists involved in biodefense research may have to wrestle with the elementary problem of retaining authority over the terms of openness for their projects. The policies, laws, and practices regarding classified and sensitive biodefense research can be interpreted in various ways and are sure to take years to evolve. In its report, the National Research Council committee exhorted, "Given the increased investments in biodefense research in the United States, it is imperative that the United States conduct its legitimate defensive activities in an open and transparent manner."[36]

The two principles on which this call for openness is based are that the larger mission of science is to increase knowledge and that medicine is based on humanitarian principles that stand above nationality or other criteria. A third principle emanates from public health, namely, that the people deserve full, clear, and accurate information on disease risks and their protective options. On the other hand, since the goal of biodefense research is explicitly for national security, US military, intelligence, and homeland security agencies may legitimately impose secrecy on biological research or restrain the publication of its results. This restriction would be analogous to those already imposed on government scientists involved in nuclear and encryption research, which are integral to weapons programs. It is highly inadvisable that biology should become part of the defense industry, that is, have its Los Alamos or other secret government research enclave, because its goals are medical.

The international character of American science and medicine also stands in potential conflict with narrowly defined goals of national security. American biologists collaborate with colleagues worldwide and frequently attend and present their research at international meetings. Would this information sharing be curtailed by the US government? Within the United States, the biological sciences are international in their education of and reliance on foreign talent and expertise. A third of all US science degrees and half of the engineering degrees each year are awarded to foreign students, many of whom remain in the US workforce.[37] US academic medicine has for decades been a training ground for graduates of overseas medical schools, who make up a third or more of public health and urban hospital staffs. In 2003, half the technicians at NIH were noncitizens.[38]

In the aftermath of September 11 and the anthrax letters, federal concerns about educating foreigners in the sciences gradually took shape. President Bush's October 2002 Presidential Decision Directive (PDD-2), for example, allows the US government to "prohibit certain international students from receiving education and training in sensitive areas, including areas of study with direct application to the development and use of Weapons of Mass Destruction." The intent of the directive was the protective restriction of dangerous knowledge, but the determination of which students might be national security threats and the definition of the term "sensitive areas" may take years to clarify.

As with the 2003 smallpox vaccination campaign, large biodefense initiatives that appear risky to the public may have to be scaled down. Plans to increase Biosafety Level 4 containment laboratories, initially ambitious, have encountered resistance from local communities in Davis, California, Boston, Massachusetts, and Galveston, Texas, from those fearful of Sverdlovsk-type aerosol accidents or other disease disasters.

Biodefense research in general could have an important deterrent effect, in that no adversary would bother with an agent against which Americans might be well protected. But there are so many agents that the problem is hydra-headed. Biodefense research and Level 4 laboratories unfortunately have the potential for increasing risks of exposure to the public. Therefore, the scale and substance of the entire enterprise needs careful, ongoing assessment. More research on select agents increases the known reserves of dangerous pathogens and therefore amplifies the risks of accidental exposure to laboratory workers as well as the risks of theft or biocrimes. At NIAID, which dispenses its research funding nationwide, biodefense research started with standard biological weapons agents, such as those for anthrax, smallpox, tularemia, and viral hemorrhagic fevers. Research can extend to dozens more agents and to genetically engineered pathogens.

NIAID director Anthony Fauci has advised that a whole new generation of biologists should be engaged in the field of biodefense research. But increased numbers of scientists trained to work specifically with biological agents may in itself pose a threat to civilian safety. As discoveries move from the laboratory to final standardized products, more scientists and technicians will gain experience in animal and human testing of new vaccines and drugs and other products. This experience would presumably include tests and trials that simulate aerosol exposure during a biological weapons attack. Yet biodefense expansion proceeded with little consideration by scientists of its risks or why, after fifteen years of government warnings of bioterrorist attacks, none had occurred. Apparently federal funding for technological solutions to bioterrorism largely precluded political analysis and action.

Conclusion

When restraints on biological weapons break down, the risks of unprecedented danger increase. If history is our guide, the threat of biological weapons increases in direct proportion to government secrecy, closed military cultures, and a subsequent lack of accountability to the public.

Looking back at the last century, we can feel optimistic at the progress made thus far in deflecting the risks of biological weapons. The French returned to a biological warfare program after World War II and then abandoned it. The British left off their program in the late 1950s and became leaders in treaty negotiations. The horrific program of the Japanese Army has had most of its secrets divulged and remains, with the Holocaust, one of the most distressing examples of barbarism within a great culture. The US offensive program is no more; the giant Horton Test Sphere at Fort Detrick is on the National Register as a Maryland tourist site. The totalitarian Soviet Union, a political mastodon, took its offensive program to its historical grave. Now international consensus against biological weapons can be weighed against the complexities of the new wars born of globalization, poverty, and ethnic and religious fanaticism. In all, the strengthening of the Biological Weapons Convention offers the single best hope of global restraints.

As before, the best protection against the proliferation of biological weapons is transparency. Biological weapons are indeed a different class of threat, one that is variable in scale, in source, in its many potential agents, and in its possible impact on communities, including psychological and social disruption as well as illness and death. A radical argument can be made regarding both biological weapons and bioterrorism: that the policy goal should be an unobstructed flow of reliable information, in public education, in the review of scientific research, in disease reporting and diagnosis, in forensic investigation, in intelligence, in government oversight, and in international treaty compliance measures. Military and intelligence agencies may balk and pharmaceutical corporations may protest. It is for government leaders to show that their first priority is the protection of public lives, without qualification.

By virtue of its power and its democratic tradition, the United States bears a special responsibility to promote comprehensive, long-term restraints against biological weapons. Government secrecy, as the twentieth century shows, caused a degradation of biology and increased mortal risk to unsuspecting civilian targets. The antidote to a reoccurrence of programs and threats worse than the ones of the past is the fostering of international relations that increase mutual trust and reduce the drastic economic inequalities that divide the world and obstruct communication.

The United States is uniquely suited to lead this mission, if it can reconfigure its role in a politically complex world where national defense requires international cooperation. In all matters of disease transmission, freely shared information is essential to our protection. As Senator Daniel Patrick Moynihan wrote in 1998, commenting on the end of years of Cold War secrecy and weapons proliferation, "Openness is now a singular, and singularly American advantage."[39] It is an advantage that can be too easily lost, in short-sighted homeland security policies and in failed relations with other nations, to our peril and the peril of future generations. In military history, biological weapons were a failed innovation. May they remain so.

NOTES

PREFACE

1. Frederic J. Brown, *Chemical Warfare: A Study in Restraints* (Princeton, NJ: Princeton University Press, 1968), 290–316; Stockholm International Peace Research Institute (SIPRI), *The Problem of Chemical and Biological Warfare*, vol. 1, *The Rise of CB Weapons* (New York: Humanities Press, 1971), 294–335.
2. Thomas C. Schelling, *The Strategy of Conflict* (Cambridge, MA: Harvard University Press, 1960), 231.
3. Ulrich Beck, *Risk Society: Towards a New Modernity* (London: Sage, 1992).

INTRODUCTION

1. Mark Wheelis, "Biological Warfare Before 1914," in Erhard Geissler and John Ellis van Courtland Moon, eds., *Biological and Toxin Weapons: Research, Development and Use from the Middle Ages to 1945* (New York: Oxford University Press, 1999), 8–34; Elizabeth A. Fenn, *Pox Americana: The Great Smallpox Epidemic of 1775–82* (New York: Hill and Wang, 2001), 88–91.
2. For details on early chemical and biological treaties, see Stockholm International Peace Research Institute (SIPRI), *The Problem of Chemical and Biological Warfare*, vol. 3, *Chemical and Biological Warfare and the Law of War* (New York: Humanities Press, 1973), and vol. 4, *Chemical and Biological Disarmament Negotiations, 1920–1970* (New York: Humanities Press, 1971). The principal author of this six-volume series is Julian Perry Robinson.
3. Frits Kalshoven, "Arms, Armaments, and International Law," *Hague Academy of International Law* 191 (1985): 191–339.
4. Guilio Douhet, *Command of the Air* (London: Faber & Faber, 1942); Alfred F. Hurley, *Billy Mitchell: Crusader for Air Power* (Bloomington: Indiana University Press, 1975); see also John Buckley, *Air Power in the Age of Total War* (Bloomington: Indiana University Press, 1999).

5. See John Keegan, *The Face of Battle* (New York: Viking, 1977), for the historic overview of how weapons technology distances the attacker from the attacked.
6. This famous phrase is attributed to Fritz Haber, a pioneer of gas warfare who received the 1918 Nobel Prize in chemistry.
7. See Brian Balmer, *Britain and Biological Warfare: Expert Advice and Science Policy, 1930–65* (New York: Palgrave, 2001), 51–52.
8. Arthur Marwick, *Britain in the Century of Total War: War, Peace and Social Change* (Boston: Little, Brown, 1968).
9. Balmer, *Britain and Biological Warfare*, 72–73.
10. Frederic J. Brown, *Chemical Warfare: A Study in Restraints* (Princeton, NJ: Princeton University Press, 1968); SIPRI, *The Problem of Chemical and Biological Warfare*, vol. 1, *The Rise of CB Weapons* (New York: Humanities Press, 1971), 294–335. For the classic organizational analysis of how the use of nuclear weapons was avoided, see Graham T. Allison and Philip Zelikow, *Explaining the Cuban Missile Crisis*, 2d ed. (Glenview, IL: Longman, 1999).
11. See Barry Posen, *The Sources of Military Doctrine* (Ithaca, NY: Cornell University Press, 1984), 13.
12. See Thomas C. Schelling, *The Strategy of Conflict* (Cambridge, MA: Harvard University Press, 1980), 208–212.
13. Max Weber, "Bureaucracy," in H.H. Gerth and C. Wright Mills, eds., *Essays in Sociology* (New York: Oxford University Press, 1946), 230–243; Aaron Wildavsky, "The Self-Evaluating Organization," *Public Administration Review* 32 (September–October 1972), 509–520.
14. Rita R. Colwell and Raymond A. Zilinskas, "Bioethics and the Prevention of Biological Warfare," in Raymond A. Zilinskas, ed., *Biological Warfare: Modern Offense and Defense* (London: Lynne Rienner, 2000), 225–245.
15. Hugh Gusterson, *Nuclear Rites: An Anthropologist Among Weapons Scientists* (Berkeley: University of California Press, 1996), 40–43.
16. See Ken Alibek with Stephen Handelman, *Biohazard: The Chilling True Story of the Largest Covert Biological Weapons Program in the World* (New York: Random House, 1999), 262.
17. France reinitiated its program after World War II. It acceded to the Biological Weapons Convention in 1984.
18. Mary Kaldor, *New and Old Wars: Organized Violence in a Global Era* (Cambridge, MA: Polity Press, 1999), 2–3; on regional conflict, see also Edward N. Luttwak, "Towards Post-heroic Warfare," *Foreign Affairs* 74, no. 3 (1995): 109–123.
19. Michael Walzer, *Just and Unjust Wars: A Moral Argument with Historical Illustrations* (New York: Basic Books, 1977), 197–206.
20. Matthew Meselson, "Bioterrorism: What Can Be Done?" in Robert B. Silvers and Barbara Epstein, eds., *Striking Terror: America's New War* (New York: New York Review Books, 2002), 257–276.

1. Biological Agents and Disease Transmission

1. See Robert S. Desowitz, *The Malaria Capers: More Tales of Parasites and People, Research, and Reality* (New York: Norton, 1991), for a popular overview of the history of "miasma" and malaria research; see also Gordon Harrison, *Mosquitoes, Malaria, and Man: A History of the Hostilities Since 1880* (London: J. Murray, 1978); Alain Corbin, *The Foul and the Fragrant: Odour and the French Social Imagination* (Oxford: Berg, 1986); on fatalism, see Yi-Fu Tuan, *Landscapes of Fear* (Minneapolis: University of Minnesota Press, 1982), 88–104.
2. Mark Wheelis, "Biological Sabotage in World War I," in Erhard Geissler and John Ellis van Courtland Moon, eds., *Biological and Toxin Weapons: Research, Development and Use from the Middle Ages to 1945* (New York: Oxford University Press, 1999), 35–62.
3. Ibid., 62.
4. Theodor Rosebury and Elvin A. Kabat, with the assistance of Martin H. Boldt, "Bacterial Warfare," *Journal of Immunology* 56, no. 1 (1947): 7–96.
5. For background on the impact of medical science since 1700, see William H. McNeill, *Plagues and Peoples* (New York: Doubleday, 1977), 208–257. For broad historical overview, see Jared Diamond, *Guns, Germs, and Steel: The Fates of Human Societies* (New York: Norton, 1999).
6. Norman Longmate, *King Cholera: The Biography of Disease* (London: H. Hamilton, 1966), 1–5.
7. Joseph Robins, *The Miasma: Epidemic and Panic in Nineteenth Century Ireland* (Dublin: Institute of Public Administration, 1995), 62–110.
8. Ibid., 9–31.
9. Isobel Grundy, *Lady Mary Wortley Montagu: Comet of the Enlightenment* (New York: Oxford University Press, 1999); McNeill, *Plagues and People*, 223–227.
10. Olivier Lepick, "French Activities Related to Biological Warfare," in Geissler and Moon, *Biological and Toxin Weapons*, 70–90. Lepick's article gives an overview of the French program until 1940.
11. Wheelis, "Biological Sabotage," 56.
12. Lepick, "*French Activities*," 72; Erhard Geissler "Biological Warfare Activities in Germany, 1923–45," in Geissler and Moon, *Biological and Toxin Weapons*, 91–126. Although German intelligence procured a secret French document on anthrax spores as a biological weapons agent. German medical and veterinary experts who read it were skeptical about their military usefulness.
13. Quoted and summarized in Lepick, "French Activities," 76–78.
14. Lepick, "French Activities," 77.
15. Quoted in Sir Norman Angell, *The Menace to Our National Defence* (London: Hamish Hamilton, 1934), 61. Angell reviews air retaliation as defense in total war.
16. Throughout the 1930s, influenced by Trillat, French biologists, among others, felt that dried agents were nearly useless. See R. DuPérié, "La guerre bactérienne," *Jour-*

nal de Médecine de Bordeaux (10 January 1935): 7–9; Médecin-Colonel Le Bordellès, "La guerre bactériologique de la défense passive antimicrobienne," *Le Bulletin Médical* 53, no. 10 (1939): 179–185; Giannino de Cesare, "La guerra microbia," *Minerva Medica* 26 (1935): 733–737.

17. Major Leon A. Fox, "Bacterial Warfare: The Use of Biologic Agents in Warfare" *The Military Surgeon* 72, no. 3 (1933): 189–207. Fox writes, "No living organism will withstand the temperature generated by an exploding artillery shell" (205).
18. See A. Rochaix, "Epidémies provoquées: A propos de la guerre bactérienne" *Revue d'Hygiène* 58, no. 3 (1936): 161–180, for summaries of Trillat. Rochaix notes that Louis Pasteur might have been first to have the idea of provoking epidemics, in 1887, when he successfully used chicken cholera to exterminate rabbits that threatened the Pommery champagne cellar in Rheims. The next year Pasteur sent one of his scientists to Australia, but the disease proved less effective and workers, fearful of contagion, mounted a campaign against the Pasteur process (169–170).
19. Andy Thomas, *Effects of Chemical Warfare: A Selective Review and Bibliography of British State Papers* (London: Taylor & Francis, 1985), 49.
20. R.S. Young, "The Use and Abuse of Fear: France and the Air Menace in the 1930's," *Intelligence and National Security* 2, no. 4 (1987): 88.
21. Dupérié, "La guerre bactérienne," 9; see Rochaix, "Epidémies provoquées," 174.
22. Jean Duffour, "La guerre bactériologique," *Journal de Médecine de Bordeaux* (16–23 October 1937): 333–349.
23. Le Bordellès, "La guerre bactériologique," 184–185.
24. Fox, "Bacterial Warfare," 207.
25. Rosebury and Kabat, "Bacterial Warfare." For general reference, see C. A. Mims, *The Pathogenesis of Infectious Disease,* 3d ed. (New York: Academic Press, 1987). See also Frederick R. Sidell, Ernest T. Takafuji, and David R. Franz, eds., *Medical Aspects of Chemical and Biological Warfare* (Washington, DC: Office of the Surgeon General, Department of the Army, 1997), 437–685. These pages cover general history, medicine and principal antihuman agents; and David R. Franz, Peter B. Jahrling, Arthur M. Friedlander, et al., "Clinical Recognition and Management of Patients Exposed to Biological Agents," in Joshua Lederberg, ed., *Biological Weapons: Limiting the Threat* (Cambridge, MA: MIT Press, 1999), 37–80.
26. Rosebury and Kabat, "Bacterial Warfare," 20.
27. Rosebury and Kabat knew about the Chinese accusations that Japan had used plague-infected fleas against them. See chapter 4 of this book.
28. Quoted in Robert Harris and Jeremy Paxman, *A Higher Form of Killing: The Secret Story of Chemical and Biological Warfare* (New York: Hill and Wang, 1982), 69.
29. Rosebury and Kabat, "Bacterial Warfare," 11.
30. Ibid., 10.
31. For comparison, see World Health Organization, *Health Aspects of Chemical and Biological Weapons* (Geneva: World Health Organization, 1970), 60–83.
32. Rosebury and Kabat, "Bacterial Warfare," 42.
33. See H.N. Glassman, "Discussion," *Bacteriologic Review* 30 (1966): 657–659; Matthew Meselson, Jeanne Guillemin, Martin Hugh-Jones, A. Langmuir, Ilona

Popova, Alexis Shelokov, and Olga Yampolskaya, "The Sverdlovsk Anthrax Outbreak of 1979," *Science* 266 (518): 1202–1208; Jeanne Guillemin, *Anthrax: The Investigation of a Deadly Outbreak* (Berkeley: University of California Press, 1999), 278–279.

34. Rosebury and Kabat, "Bacterial Warfare," 45.
35. Ibid., 14.
36. Ibid., 31.
37. John Duffy, *The Sanitarians: A History of American Public Health* (Urbana: University of Illinois Press, 1990), 256–264.
38. Theodor Rosebury, *Peace or Pestilence? Biological Warfare and How to Avoid It* (New York: McGraw-Hill, 1949).

2. The United Kingdom and Biological Warfare

1. Barry R. Posen, *Inadvertent Escalation: Conventional War and Nuclear Risks* (Ithaca, NY: Cornell University Press, 1992), 20. In his discussion of determinants of escalation, Posen applies the idea of Carl von Clausewitz, *On War* (Princeton, NJ: Princeton University Press, 1976), 117–118, concerning the contradictory, false, and uncertain intelligence in war, often referred to as the "fog of war," and the difficulties faulty information poses for military decision makers.
2. H. Wickham Steed, "Aerial Warfare: Secret German Warfare," *The Nineteenth Century and After* 116 (July 1934): 1–15; "The Future of Warfare," *The Nineteenth Century and After* 116 (August 1934): 129–140.
3. Steed, "Aerial Warfare," 1.
4. See critical comments by Ernst Burkhardt under "Correspondence," *The Nineteenth Century and After* 116 (September 1934): 331–336; and Wickham Steed's rejoinder, ibid., 337–339. See also Martin Hugh-Jones, "Wickham Steed and German Biological Warfare Research," *Intelligence and National Security* 7, no. 4 (1992): 380–402.
5. Brian Balmer, *Britain and Biological Warfare: Expert Advice and Science Policy, 1930–65* (London: Palgrave, 2001), 27.
6. Robert Harris and Jeremy Paxman, *A Higher Form of Killing: The Secret Story of Chemical and Biological Warfare* (New York: Hill and Wang, 1982), 65–67, 135.
7. Erhard Geissler, "Biological Warfare Activities in Germany, 1923–45," in Erhard Geissler and John Ellis van Courtland Moon, eds., *Biological and Toxin Weapons: Research, Development and Use from the Middle Ages to 1945* (New York: Oxford University Press, 1999), 91–126.
8. Stephen Roskill, *Hankey: Man of Secrets,* vol. 3, *1931–1963* (London: Collins, 1974).
9. It is worth noting that the UK program was technically prepared to start research on biological weapons and might have done so eventually, perhaps with initiatives from its chemical corps or that of the United States.
10. See Gradon B. Carter and Graham Pearson (both formerly at Porton Down), "British Biological Warfare and Biological Defence, 1925–45," in Geissler and Moon, *Biological and Toxin Weapons,* 168–189. "Hankey, later Lord Hankey, must be regarded as the founding father of biological warfare and biological warfare defence in the UK."

11. Balmer, *Britain and Biological Warfare*, 29–54.
12. PRO WO188/650 CBW1 CID Subcommittee on biological warfare (2 November 1936).
13. Gradon B. Carter and Graham S. Pearson, "British Biological Warfare and Biological Defence, 1925–45," in Geissler and Moon, *Biological and Toxin Weapons*, 168–189. Martin Hugh-Jones ("Wickham Steed," 393) observed Roskill's note that while in military service Hankey had his experience of epidemics as well.
14. Balmer, *Britain and Biological Warfare*, 14–28.
15. Arthur Marwick, *Britain in the Century of Total War: War, Peace and Social Change* (Boston: Little, Brown, 1968), 266.
16. See John Bryden, *Deadly Allies: Canada's Secret War 1937–1947* (Toronto: McClelland & Stewart, 1989), 34–57; Donald Avery, "Canadian Biological and Toxin Warfare," in Geissler and Moon, *Biological and Toxin Weapons*, 190–214.
17. See Lloyd Stevenson, *Sir Frederick Banting* (Toronto: Ryerson Press, 1946), 151–163; and Paul De Kruif's more dramatic version, "Banting: Who Found Insulin," in *Men Against Death* (New York: Harcourt, Brace, 1932), 59–87.
18. Frederick O. Banting, "Memorandum on the Present Situation Regarding Bacterial Weapons" (undated), PRO, WO188/653 10. This statement was brought by Banting to England in November 1939.
19. Banting, "Memorandum," 9.
20. Ibid., 3.
21. Ibid., 8.
22. Quoted in Bryden, *Deadly Allies*, 38.
23. Roskill, *Hankey*, 467–468.
24. PRO, CAB104/234. Letter, Hankey to Group Captain Elliot (26 July 1940). See also Balmer, *Britain and Biological Warfare*, 29–54.
25. John Buckley, *Air Power in the Age of Total War* (Bloomington: Indiana University Press, 1999), 81.
26. Balmer, *Britain and Biological Warfare*, 36–37.
27. Winston Churchill, *The Gathering Storm* (Boston: Little, Brown, 1948), 34.
28. Porton scientists have deemphasized the potential impact the cattle cakes would have had on civilians. See Peter H. Hammond and Gradon B. Carter, *From Biological Weapons to Health Care: Porton Down, 1940–2000* (New York: Palgrave, 2002), 12–13. Carter and Pearson, "British Biological Warfare," 181–183, describe the production and planned delivery system for the cattle cakes.
29. Bryden, *Deadly Allies*, 40–43.
30. Ibid., 50.
31. Quoted in Harris and Paxman, *Higher Form*, 69. This was the quote that Rosebury and Kabat refer to in their 1942 report.
32. Balmer, *Britain and Biological Warfare*, 45–46.
33. J. B. S. Haldane, *Callinicus: A Defence of Chemical Warfare* (London: Kegan, Paul, Trench, Trubner, 1925), 32.
34. PRO, WO188/654 BIO/5293. Fildes Notes on Professor Murray's Memorandum (7 September 1944).

3. The United States in World War II

1. Brian Balmer, *Britain and Biological Warfare: Expert Advice and Science Policy, 1930–65* (London: Palgrave, 2001), 188, for a diagram of the British 1947 committee oversight of biological weapons, brought forward from the war years.
2. PRO, WO 188/654, Letter, Professor Newitt to Fildes, 12 February 1942.
3. Secretary of State Henry L. Stimson to Secretary of State Cordell Hull, 18 February 1942, National Archives of the USA, Records of the War Department, General and Special Staffs, RG 165.
4. Robert Harris and Jeremy Paxman, *A Higher Form of Killing: The Secret Story of Chemical and Biological Warfare* (New York: Hill and Wang, 1982), 95–96. See details on Stimson's role in Barton J. Bernstein, "Origins of the Biological Warfare Program," in Susan Wright, ed., *Preventing a Biological Arms Race* (Cambridge, MA: MIT Press, 1990), 9–25. See also Bernstein's previous article, "America's Biological Warfare Program in the Second World War," *Journal of Strategic Studies* 11 (September 1988): 292–317; and the overview in John Ellis van Courtland Moon, "US Biological Warfare Planning and Preparedness: the Dilemmas of Policy," in Erhard Geissler and John Ellis van Courtland Moon, eds., *Biological and Toxin Weapons: Research, Development and Use from the Middle Ages to 1945* (New York: Oxford University Press, 1999), 217–254.
5. Harris and Paxman, *A Higher Form of Killing,* 115–118.
6. Ibid., 118.
7. June 8, 1943, statement, quoted in Frederic J. Brown, *Chemical Warfare: A Study in Restraints* (Princeton, NJ: Princeton University Press, 1968), 264. Bernstein ("Origins," 16) characterizes Roosevelt's attitude as more disengaged, so that he bequeathed to President Harry Truman "an ambiguous legacy regarding biological warfare."
8. William D. Leahy, *I Was There* (New York: McGraw-Hill, 1950), 440. Leahy was also against nuclear weapons: "These new concepts of 'total war' are basically distasteful to the soldier and sailor of my generation. Employment of the atomic bomb in war will take us back in cruelty toward noncombatants to the days of Genghis Khan" (441–442).
9. Leo P. Brophy, Wyndham D. Miles, and Rexmond C. Cochrane, *The Chemical Warfare Service: From Laboratory to Field* (Washington, DC: Office of the Chief of Military History, U.S. Army, 1959), 103.
10. Rexmond C. Cochrane, *Biological Warfare Research in the United States,* vol. II, part D, XXIII: *Chemical Plant Growth: History of the Chemical Warfare Service in World War II (1 July 1940–15 August 1945)* (Washington, DC: Office of Chief, Chemical Corps, 1947), 19. See "Organization of WRS" flow chart.
11. On accelerated vaccine development during the war, also related to biological weapons, see Kendall L. Hoyt, "The Role of Military-Industrial Relations in the History of Vaccine Innovation," PhD dissertation, MIT, 2002.
12. John Bryden, *Deadly Allies: Canada's Secret War 1937–1947* (Toronto: McClelland & Stewart, 1989), 81–84.

13. Ibid., 121.
14. See Oram C. Woolpert, letter of 24 February 1948, appended to Cochrane, *Biological Warfare Research in the United States.*
15. Brophy, Miles, and Cochrane, *Chemical Warfare Service,* 111.
16. See Woolpert, letter of 24 February 1948, which limits the tested agents to anthrax and plague.
17. See Theodor Rosebury, *Experimental Air-Borne Infection* (Baltimore: Williams and Wilkins, 1947), 48–56.
18. Harris and Paxman, *A Higher Form of Killing,* 103.
19. B.H. Liddell Hart, *History of the Second World War* (New York: G. P. Putnam's Sons, 1971), 592.
20. PRO, CAB 136/1 Memo by R. G. Peck, 12 January 1944.
21. Bryden, *Deadly Allies,* 138–161.
22. PRO, PREM 3/65.
23. Ibid.
24. Some historians from Porton contend that Churchill was not in favor of biological warfare. See Gradon B. Carter and Graham S. Pearson, "British Biological Warfare and Biological Defense, 1925–45," in Geissler and Moon, eds., *Biological and Toxin Weapons,* 168–189, 181.
25. See Leo P. Brophy and George J. B. Fisher, *The Chemical Warfare Service: Organizing for War* (Washington, DC: Office of the Chief of Military History, Department of the Army, 1959).
26. Cochrane, *Biological Warfare Research in the United States,* 209–231.
27. Ibid, 234.
28. See Bryden, *Deadly Allies,* 118–119 and 197–199, for Bryden's 1989 interview with Baldwin about the Vigo plant.
29. Cochrane, *Biological Warfare Research in the United States,* 77–80.
30. Ibid., 80.
31. The contingency plans were described by Fildes in what appears to be a November 1945 memo (PRO-DEFE 2/1252. Report to the Joint Technical Warfare Committee on "Potentialities of Weapons of Biological Warfare During the Next Ten Years."), with bomber and bomb details on page 3 of the appendix. Harris and Paxman, *A Higher Form of Killing,* reason that the plans must predate the fall of Aachen in October 1944. They may reflect initial enthusiasm for the "N" bomb, an enthusiasm that faded during the summer of 1944 as the United States delayed on bomb production and as it became increasingly clear that the Germans had no biological weapons program (104).
32. Brophy, Miles, and Cochrane, *Chemical Warfare Service,* 112–115.
33. Cochrane, *Biological Warfare Research in the United States,* 479.
34. Ibid., 509.
35. PRO-WO188/188/654 "Failure of BW Policy in the European Theatre," P. Fildes, 6 June 1944.
36. Baldwin, with his mentor Dr. Fred, continued to be active in advising on biological warfare projects, especially after the war.

37. See Bryden, *Deadly Allies*, 122–127.
38. Quoted in Cochrane, *Biological Warfare Research in the United States*, 189–191, from Secretary of War Stimson and Paul V. McNutt, Administrator, Federal Security Agency, to President Roosevelt, 12 May 1944.
39. Ibid, 226.
40. Harris and Paxman, *A Higher Form of Killing*, 96.
41. Cochrane, *Biological Warfare Research in the United States*, 474–477.
42. Ibid., 478.
43. Ibid.
44. Bernstein, "Origins of the Biological Warfare Program," 17–18.

4. Secret Sharing and the Japanese Biological Weapons Program (1934–1945)

1. William M. Creasy, "Presentation to the Secretary of Defense's Ad Hoc Committee on CEBAR," 24 February 1950, Joint Chiefs of Staff Files.
2. Rexmond C. Cochrane, *Biological Warfare Research in the United States*, vol. II, part D, XXIII: *Chemical Plant Growth: History of the Chemical Warfare Service in World War II (1 July 1940–15 August 1945)* (Washington, DC: Office of Chief, Chemical Corps, 1947), 48–488.
3. Theodor Rosebury, *Experimental Air-Borne Infection* (Baltimore: Williams and Wilkins, 1947).
4. Samuel Goudsmit, *Alsos* (Los Angeles: Tomash, 1983); John Gimbel, *Science, Technology and Reparations: Exploitation and Plunder in Postwar Germany* (Stanford, CA: Stanford University Press, 1990).
5. R.W. Home and Morris F. Low, "Postwar Scientific Intelligence Missions to Japan," *Isis* 84 (1993): 527–537. Two MIT scientists led the mission: Karl Compton, president of MIT and head of Field Service at the Office of Scientific Research and Development during the war; and Edward Moreland, dean of engineering and wartime head of the National Defense Research Committee.
6. A. T. Thompson, "Report on Japanese Biological Warfare (BW) Activities, Army Service Forces, Camp Detrick, Maryland," 31 May 1946, Fort Detrick Archives, Supplement 3e.
7. See Linda Hunt, *Secret Agenda: The United States Government, Nazi Scientists, and Project Paperclip, 1944–1990* (New York: St. Martin's Press, 1991).
8. Sheldon H. Harris, *Factories of Death: Japanese Biological Warfare 1932–45 and the American Cover-up* (London: Routledge, 1994), 169–172. In 1990 Harris visited military archives at Dugway, Utah, and copied extensive documentation of both the amnesty bargain and the Japanese reports of the program.
9. US National Archives and Records Administration, 20 March 2002, press release, "IWG Report to Congress."
10. Summary of Information, Subject Ishii, Shiro, 10 January 1947, Document 41, US Army Intelligence and Security Command Archive, Fort Meade, Maryland.

11. SWNCC (State-War-Navy Coordinating Committee) 351/1, 5 March 1947. Record Group 331, Box 1434.20, Case 330. US National Archives.
12. Marius B. Jansen, *The Making of Modern Japan* (Cambridge, MA: Harvard University Press, 2000), 668–669.
13. Peter Williams and David Wallace *Unit 731: The Japanese Army's Secret of Secrets* (Sevenoaks, Kent: Hodder and Stoughton, 1989), 75–80. The colonial army often served as a stepping-stone to Tokyo government positions.
14. Ibid., 76–78, 138. The American version of this book was published by Free Press, 1990, and deleted the parts building a case for US use in the Korean War of biological weapons based on Japanese information, specifically plague-infected fleas.
15. Sheldon Harris, "Japanese Medical Atrocities in World War II: Unit 731 Was Not an Isolated Aberration," International Citizens Forum on War Crimes and Redress, Tokyo, Japan, 11 December 1999.
16. Thompson, "Report," 1946.
17. Williams and Wallace, *Unit 731,* 207.
18. Harris, *Factories of Death,* 203.
19. One of those who received immunity in 1948, Naito Ryiochi, later founded the multinational pharmaceutical company Green Cross Corporation, became a member of the New York Academy of Sciences and received Japan's Order of the Rising Sun in 1982.
20. *Materials on the Trial of Former Servicemen of the Japanese Army Charged with Manufacturing and Employing Bacteriological Weapons* (Moscow: Foreign Language Press, 1950).
21. See chapter 7.
22. Louise Young, *Japan's Total Empire: Manchuria and the Culture of Wartime Imperialism* (Berkeley: University of California Press, 1998), 55–114.
23. Jansen, *Making of Modern Japan,* 587.
24. Muto Sanji, Alessandro Rossi, Kishi Nobusuke, Giovanni Agnelli, and Ayukawa Gisuke, "The Birth of Corporatism," in Richard J. Samuels, ed., *Machiavelli's Children* (Ithaca, NY: Cornell University Press, 2003), 124–151.
25. See Jansen, *Making of Modern Japan,* 566–624.
26. Williams and Wallace, *Unit 731,* 223.
27. Harris, *Factories of Death,* 18.
28. Williams and Wallace, *Unit 731,* 7.
29. Kei-ichi Tsuneishi, "C. Koizumi: As a Promoter of the Ministry of Health and Welfare and an Originator of the BCW Research Program," *Historia Scientiarum* 26 (1984): 95–113.
30. Thompson, "Report," 1946.
31. Kei-ichi Tsuneishi, "Research Guarded by Military Secrecy," *Historia Scientiarum* 30 (1986): 79–92.
32. "Japanese Attempts at Bacteriological Warfare in China," 9 July 1942, PRO, PREM 3/65.

33. Ishii was probably behind unsuccessful attempts to spread cholera (via dropped glass vials) in Burma. In 1944, a team of seventeen officers trained by him was sent by ship to cover the Saipan airstrip with plague-infected fleas. En route, an American submarine sank the ship. See Williams and Wallace, *Unit 731*, 81, based on correspondence with Professor Tsuneishi.
34. "Former Japanese Army Targeted Pacific for Germ Bombs," Kyodo News Service, 28 November 1993. The article cites university researchers as its source.
35. Tsuneishi, "C. Koizumi," 104.
36. Norbert E. Fell, Chief, PP-E (Planning Pilot-Engineering) Division, to Assistant Chief of Staff, G-2, GHQ, Far East Command; through Technical Director, Camp Detrick, Commanding Officer, Camp Detrick, 24 June 1947, 11.
37. Edwin V. Hill, M.D., to Gen. Alden C. Waitt, Chief Chemical Corps, Pentagon, Washington, D.C. (12 December 1947).
38. Fell to Assistant Chief of Staff, 10.
39. Martin Furmanski and Sheldon H. Harris conclude that the reports were written up by Ishikawa Tachjio, who later became president of the Kanazawa University School of Medicine; "Effects of Unsophisticated Anthrax Bioweapons I: Clinical and Pathologic Features of Victims: Evidence from Japanese Crimes against Humanity," unpublished manuscript, October 2001, 4. Of less interest were six cases of gastrointestinal anthrax and one of a death from cutaneous anthrax.
40. See Benno Müller-Hill, *Murderous Science: The Elimination by Scientific Selection of Jews, Gypsies, and Others, Germany 1933–1945* (New York: Oxford University Press, 1988); Robert Jay Lifton, *Nazi Doctors: Medical Killing and the Psychology of Genocide* (New York: Basic Books, 1986); and Robert Proctor, *Racial Hygiene: Medicine Under the Nazis* (Cambridge, MA: Harvard University Press, 1988).
41. Aaron Epstein, "MD: U.S. Hid Japan's Experiments on POWs," *Miami Herald* 7 December 1985.
42. Fell to Assistant Chief of Staff, 2.
43. Harris, *Factories of Death*, 222.
44. Hill to Waitt.
45. The *Post* article, by John Saar, was "WWII Germ Deaths Laid to Japan," 19 November 1976.
46. Robert Gomer, John W. Powell, and Bert V. A. Röling, "Japan's Biological Weapons: 1930–1945," *The Bulletin of the Atomic Scientists*, October 1981, 43–53. The substance of this article appeared under Powell's name in *Journal of Concerned Asian Scholars* 12, no. 4 (1980): 2–17.
47. Tokyo: Kai-nei-sha Publishers. See also Kei'ichi Tsuneishi and Tomizo Asano, *The Bacteriological Warfare Unit and the Suicide of Two Physicians* (Tokyo: Shincho-Sha, 1982). Tsuneishi, "Research," includes the summary list of autopsy cases obtained from the Japanese as part of the immunity bargain.
48. Williams and Wallace, *Unit 731*, 189.

49. Tamura Yoshio, "Unit 731," in Haruko Taya Cook and Theodore F. Cook, *Japan at War: An Oral History* (New York: New Press, 1992), 158–167.
50. Harris, *Factories of Death,* 152.
51. Reported in the *CBW Bulletin* 50 (December 2000): 38.
52. Reported at the Sixth International Conference on Sino-Japanese Relations (Foster City, CA, March 15, 2000) by Li Yaquan of China, secretary general of the society for the study of Japan's fourteen-year rule in Manchuria.
53. John W. Powell, "Japan's Germ Warfare: The U.S. Cover-up of a War Crime," *Bulletin of Concerned Asian Scholars* 12, no. 4 (1980): 2–17.

5. Aiming for Nuclear Scale

1. William M. Creasy, "Presentation to the Secretary of Defense's Ad Hoc Committee on CEBAR," 24 February 1950, Joint Chiefs of Staff Files.
2. Seymour Hersh, *Chemical and Biological Warfare: America's Hidden Arsenal* (Indianapolis: Bobbs-Merrill, 1968), 202.
3. Report by the Joint Strategic Plans Committee to the Joint Chiefs of Staff on Statements of Policy and Directives on Biological Warfare, JCS 1837/34, 11 June 1952.
4. Ibid., 289.
5. Ibid., 324.
6. Theodor Rosebury, *Peace or Pestilence? Biological Warfare and How to Avoid It* (New York: McGraw-Hill, 1949), 50. Rosebury details the other media coverage noted here.
7. See Brian Balmer, *Britain and Biological Warfare: Expert Advice and Science Policy, 1930–65* (London: Palgrave, 2001), 70–73, for discussion of the US and UK intelligence evaluations of the Soviet postwar biological threat. Balmer notes, "In the vacuum left by such reports, assessments were frequently based on the assumption that UK and US possibilities reflected Russian capabilities" (72).
8. Rosebury, *Peace or Pestilence,* 150.
9. Study by the Joint Advanced Study Committee on Biological Warfare. JCS, 1837/26, 7 September 1951, 290.
10. Rexmond C. Cochrane, *Biological Warfare Research in the United States,* vol. II, part D, XXIII: *Chemical Plant Growth: History of the Chemical Warfare Service in World War II (1 July 1940–15 August 1945)* (Washington, DC: Office of Chief, Chemical Corps, 1947), 508–512.
11. Hersh, *Chemical and Biological Warfare,* 133–138.
12. Over the decades, the MRD and the MRE gradually extended animal research on inhalational infections beyond anthrax and plague to tuberculosis, brucellosis, tularemia, Venezuelan equine encephalitis virus, rabbitpox virus, influenza, Marburg virus, Ebola virus, Vaccinia virus, and smallpox virus, among other pathogens. See Peter Hammond and Gradon Carter, *From Biological Warfare to Healthcare: Porton Down, 1940–2000* (New York: Palgrave, 2002), 65–66.
13. Hammond and Carter, *From Biological Warfare to Healthcare,* 76–92
14. Ibid., 66.

15. Balmer, *Britain and Biological Warfare*, 147.
16. P. C. B. Turnbull "Anthrax Vaccines: Past, Present, and Future" *Vaccine* 9 (August 1991): 533–539. After several years, Harry Smith left Porton for an academic position.
17. Hammond and Carter, *From Biological Warfare to Healthcare*, 102–104; Balmer, *Britain and Biological Warfare*, 98–101.
18. O. H. Wansbrough-Jones, "Future Development of Biological Warfare" PRO, DEFE 2/1252, 7827, 4. On the same page, the author also points out that medical defenses against BW could "render it obsolete, but that in the meanwhile preparedness was recommended...our position will be bad if other countries develop an effective weapon."
19. Balmer, *Britain and Biological Warfare*, 79–85.
20. Dorothy L. Miller, "History of the Air Force Participation in Biological Warfare Program 1944–1951," Historical Study No. 194, Wright Patterson Air Force Base, Office of the Executive Air Materiel Command, September 1952, 2.
21. Miller, "History of Air Force Participation," 6.
22. The Air Force had its own Air Research and Development Command, which worked directly with CWS, and its logistics and training division, Air Materiel Command. It also had Wright Air Development Center, which made major judgments on biological-warfare projects, and the Air Force Armament Center, in charge of tests at Eglin Air Force Base in Florida and Holloman Air Force Base in New Mexico.
23. *Report of the International Scientific Commission for the Investigation of the Facts Concerning Bacterial Warfare in Korea and China* (Beijing, 1952). The reports of the interviews with the four POWs suggest highly moving performances. The four service men recanted their testimonies once they were released and returned home. See Stockholm International Peace Research Institute (SIPRI), *The Problem of Chemical and Biological Warfare*, vol. 4, 196–223.
24. Numerous articles appeared following the Japanese journalist's report. See Milton Leitenberg, "New Russian Evidence on the Korean War Biological Warfare Allegations: Background and Analysis," *Cold War International History Project Bulletin* 11 (1998): 185–199, and "The Korean War Biological Weapons Allegations: Additional Information and Disclosure," *Asian Perspectives* 24, no. 3 (2000): 159–172; John Ellis van Courtland Moon, "Dubious Allegations," *Bulletin of the Atomic Scientists* 55, no. 3 (1999): 70–72.
25. Miller, "History of Air Force Participation," 76.
26. Ibid.
27. Ibid., 79.
28. Ibid., 52–53.
29. See Erving Goffman, *Frame Analysis: An Essay on the Organization of Experience* (Cambridge, MA: Harvard University Press, 1974), 59–62. See also Trevor Pinch, "'Testing—One, Two, Three...Testing': Toward a Sociology of Testing," *Science, Technology, & Human Values* 18, no. 1 (1993): 25–41.
30. Hammond and Carter, *From Biological Warfare to Healthcare*, 21–27.
31. D. W. Henderson and J. D. Morton, "Operation Harness 1947–1949," Ministry of Supply, Biological Warfare Sub-Committee, 20 July 1949. PRO, DEFE 10/263, 4.

32. Ibid., 4–5. Problems abounded at the site: escaped monkeys, working in masks and suits at high temperatures, loading bomb fill by hand, moving sheep into dinghies, shark attacks, one case of brucellosis, and eight cases of venereal disease among personnel. In the end, as Paul Fildes and Lord Stamp pointed out, Operation Harness told little of the long-range effects of pathogens over land, where they were most likely to be used. See Balmer, *Britain and Biological Warfare*, 104–110.
33. Miller, "History of Air Force Participation," 64, writes, "In order to produce an efficient aerosol-producing munition, the general principles governing BW aerosol generation, stability, and dynamics had to be understood. Information had to be obtained about the mechanisms of aerosol formation, the physical stability of various types of aerosols under various environmental conditions, and the dynamics of aerosol dispersion and travel. It was also necessary to relate the generating mechanisms to the biological character of the agent and to its suspending medium, and to determine to what extent the agent-fill might be modified to meet the requirements of the disseminating device. Close coordination was required with other projects dealing with aerosol stability, infectivity or toxicity, and with storage stability."
34. Munition Expenditure Panel, "Preliminary Discussion of Methods for Calculating Munition Expenditure, With Special Reference to the St Jo Program," 11 August 1954, Camp Detrick, Frederick, Maryland.
35. Ibid., 6.
36. In the St Jo report (p. 71), the Salisbury tests are cited as "Diffusion of Smoke in Built-up Area: Porton Technical paper No. 193; and Wind Tunnel Experiments of Diffusion in a Built-up Area: Porton Technical Paper No. 257.
37. See Leonard A. Cole, *Clouds of Secrecy: The Army's Germ Warfare Tests Over Populated Areas* (Lanham, MD: Rowman and Littlefield, 1988).
38. The contractor was the Ralph M. Parsons Company, and the Stanford-Parsons data are cited frequently in the St Jo report.
39. Munition Expenditure Panel, "Preliminary Discussion."
40. Miller, "History of Air Force Participation," 90. Miller goes on to note: "The test program was well conceived and admirably conducted. Coordination among the services was excellent. For the first time the Chemical Corps and the Air Force knew from the beginning what the development work was meant to accomplish. All were agreed upon the basic essentials. Logistic problems were placed alongside research and development in relative importance."
41. Ibid., 194–195. See Nicholas Hahon, *Screening Studies with Variola Virus (U)* (Fort Detrick, MD: US Army Chemical Corps, Biological Warfare Laboratories, October 1958) (Control No. 57-FD-S-1620); Mary S. Watson, *Variola Virus: A Survey and Analysis of the Literature* (Fort Detrick, MD: Chemical Corps Research and Development Command, US Army Biological Laboratories, November 1960).
42. US Army, *U.S. Army Activity in the U.S. Biological Warfare Programs*, vol. 1, 24 February 1977, 125–140.
43. Munition Expenditure Panel, "Preliminary Discussion," 9.
44. Jonathan D. Moreno, *Undue Risk: Secret State Experiments on Humans* (New York: W. H. Freeman, 1999), 254–262. The cadre of physicians in charge afterward be-

came formally recognized as the US Army Medical Unit. In 1955, the Army signed a contract with Ohio State University to test tularemia on prisoners at Ohio State penitentiary. Two strains were standarized

45. P. S. Brachman, S. A. Plotkin, F. H. Bumford, et al., "An Epidemic of Inhalation Anthrax: The First in the Twentieth Century," *American Journal of Hygiene* 1960 (72): 6–23.
46. JCS 1837/34, 326–330. In this document, Secretary of Defense Robert Lovett appears more willing to forego the retaliation-only policy than the Joint Chiefs.
47. Miller, "History of Air Force Participation," 58.
48. Ibid. Here Miller refers to a longstanding argument about whether the retaliation policy had complicated operational planning and discouraged competent personnel from entering the program. As it was, the Air Force had to have a program, but "did not have to guarantee results."
49. Quoted in Raymond L. Garthoff, *Soviet Strategy in the Nuclear Age* (New York: Praeger, 1958), 104.
50. US Army, "U.S. Army Activity," chap. 4.
51. In the 1954 US Army Field Manual 27–10, *Law of Land Warfare*, it was stipulated: "Gas warfare and bacteriological warfare are employed by the United States against enemy personnel only in retaliation for their use by the enemy." Quoted in Hersh, *Chemical and Biological Warfare*, 23. In another manual, US Army Field Manual 101–40, *Armed Forces Doctrine for Chemical and Biological Weapons Employment and Defense*, presidential control is clear: "The decision for U.S. forces to use chemical and biological weapons rests with the President of the United States." Quoted in Hersh, *Chemical and Biological Warfare*, 24.
52. Ibid.
53. Balmer, *Britain and Biological Warfare*, 160–162.
54. Quoted in Hersh, *Chemical and Biological Warfare*, 28.
55. Hersh, *Chemical and Biological Warfare*, 34–35.
56. The USG updates on the release of declassified information on SHAD are available at http://deploymentlink.osd.mil/current_issues/shad.
57. John Morrison, *DTC Test 68–50: Test Report* (Fort Douglas, UT: Department of the Army, Deseret Test Center, 1969). See Ed Regis, *The Biology of Doom* (New York: Henry Holt, 1999), 200–206. See also ibid., 188–192, on the involvement of the Smithsonian Institution in the Pacific Ocean Biological Survey Program.

6. The Nixon Decision

1. Peter Hammond and Gradon Carter, *From Biological Warfare to Healthcare: Porton Down, 1940–2000* (New York: Palgrave, 2002), 120. This retreat from an offensive posture was publicly stated in 1959 by the head of the Ministry of Supply.
2. "Discussion at the 435th Meeting of the National Security Council, February 18, 2960," Dwight D. Eisenhower Library; Eisenhower Papers, 1953–61 (Ann Whitman file), 5. Two highly placed science advisors, physicist Herbert York and chemist George Kistiakowsky, expressed optimism at this meeting that US scientists could

develop chemical and biological weapons on a par with nuclear arms. Both later repudiated this earlier enthusiasm.

3. Ibid., 6.
4. *Boston Globe*, 21 October 1960.
5. Arthur M. Schlesinger Jr., *A Thousand Days* (Boston: Houghton Mifflin, 1965), 318–319.
6. Committee on Science and Astronautics, US House of Representatives, *Technical Information for Congress: A Report to the Subcommittee on Science Research, and Development* (Washington, DC: Government Printing Office, 1971), 537.
7. Forrest Frank, "U.S. Arms Control Policymaking: The 1972 Bacteriological Treaty Case," PhD dissertation, Naval War College, 1974, 35–36.
8. "Memorandum from the Acting Director of the Arms Control and Disarmament Agency" (Adrian Fisher), Washington, 14 March 1962; Kennedy Library, National Security Files, Kaysen Series, Disarmament, Basic Memoranda, 2/62–4-4/62, 387.
9. Carl Kaysen, "Memorandum for the President," White House, 5 April 1962; Kennedy Library, National Security Files, 261, ACDA, Disarmament, 18-Nation Conference, Geneva, 4/1/62–4/11/62.
10. "Memorandum of Conversation," 8 October 1963, Arms Control and Disarmament Agency; Kennedy Library, National Security Files, Disarmament, Committee of Principles, 3/61–11/63.
11. William C. Foster, "Memorandum for the Members of the Committee of Principals," 29 October 1963, Arms Control and Disarmament Agency; McGeorge Bundy, "Memorandum for the Director, Arms Control and Disarmament Agency," White House, 5 November 1963; William C. Foster, "Memorandum for the Members of the Committee of Principals," 12 November 1963, Arms Control and Disarmament Agency; Kennedy Library, National Security Files, 255. During the previous summer, Harvard biologist Matthew Meselson worked for the Arms Control and Disarmament Agency, researching biological weapons policy, and had submitted his report recommending its review to Foster.
12. Charles Mohr, "U.S. Spray Planes Destroy Rice in Viet Cong Territory," *New York Times*, 21 December 1965.
13. See Rachel Carson, *Silent Spring* (New York: Houghton Mifflin, 1994), 34–37.
14. Jean Mayer, "Crop Destruction in Vietnam," *Science* 152, no. 3720 (15 April 1966): 291.
15. Secretary Rusk Discusses the Use of Tear Gas in Vietnam," *US Department of State Bulletin* 52, no. 1346 (12 April 1965): 528–529.
16. This incident, widely covered in the press, became known as the Utter affair, after the name of the officer who ordered riot-control gas used, perhaps against certain prohibitions issued by the Pentagon. Those restrictions were later denied. See Seymour Hersh, *Chemical and Biological Warfare: America's Hidden Arsenal* (Indianapolis: Bobbs-Merrill, 1968), 174–175.
17. Department of Defense, "Riot Control Procurement Programs for Southeast Asia," in *The Geneva Protocol Hearings before the Committee on Foreign Relations, United States Senate, March 1971* (Washington, DC: US Government Printing Office, 1972), 307.

18. Neil Sheehan, "Tear Gas Dropped Before B-52 Raid," *New York Times*, 22 February 1966.
19. See Frank, "U.S. Arms Control Policymaking," 53. In April 1974 Frank interviewed an anonymous Army officer who took credit for developing the fuse and detonators to disperse CS riot control agent from dropped drums in order to increase enemy casualties.
20. For interpretation of treaty law regarding US chemical use in Vietnam, see R. R. Baxter and Thomas Buergenthal, "Legal Aspects of the Geneva Protocol of 1925," *American Journal of International Law* 64, no. 4 (1970): 853–879.
21. Stockholm International Peace Research Institute (SIPRI), *The Problem of Chemical and Biological Warfare*, vol. 4, *CB Disarmament Negotiations, 1920–1970* (New York: Humanities Press, 1971), 243–247. Italy and Egypt ratified the protocol in 1928.
22. P. M. Haas, "Introduction: Epistemic Communities and International Policy Coordination," *International Organization* 46, no. 1, (1992): 1–35.
23. For a contemporary summary and analysis of this activity, see Charles Ruttenberg, "Political Behavior of American Scientists: The Movement Against Chemical and Biological Warfare," PhD dissertation, New York University, 1972.
24. "FAS Asks Clarification of U.S. Policies on Germ Warfare," *Bulletin of the Atomic Scientists* 8 (June 1952): 130.
25. Pugwash was named after the Nova Scotia village where the group's first supporter, Cyrus Eaton, had a summer estate used for conferences.
26. "'On Chemical and Biological Warfare': Statement of the Fifth Pugwash Conference," *Bulletin of the Atomic Scientists* 15 (October 1959): 337–339. See Julian Perry Robinson, "The Impact of Pugwash on the Debates of Chemical and Biological Weapons," *Annals of the New York Academy of Sciences* 866 (30 December 1998): 224–252.
27. SIPRI, *The Problem of Chemical and Biological Warfare*, 339. See also Joseph Rotblat, *Pugwash: The First Ten Years* (New York: Humanities Press, 1968).
28. Rotblat, *Pugwash*, 206.
29. Supported by the Swedish government, SIPRI was established in 1966 to study and publish on questions of arms control and disarmament.
30. SIPRI, *The Problem of Chemical and Biological Weapons*, vols. 1–6. Individual volumes are cited throughout this text. Julian Robinson, later codirector of the Harvard Sussex Program on CBW Armament and Arms Limitation, was the author of the indispensable first two volumes in this series.
31. "Letter to the President," *Bioscience* 17 (January 1967): 10.
32. These and other advocates are represented in Steven Rose, ed., *CBW: Chemical and Biological Warfare* (Boston: Beacon Press, 1968).
33. Theodor Rosebury, *Peace or Pestilence? Biological Warfare and How to Avoid It* (New York: McGraw-Hill, 1949).
34. Joel Primack and Frank von Hippel, "Matthew Meselson and Federal Policy on Chemical and Biological Warfare," in *Advice and Dissent: Scientists in the Political Arena* (New York: Basic Books, 1974), 143–164.
35. The petition was initially released to the press with the signatures of twenty-two leading scientists, including seven Nobel Prize winners. With the assistance of biochemist

Milton Leitenberg, who later devoted his career to the study of chemical and biological arms control, the petition was open to all US scientists with doctoral degrees.

36. Foy D. Kohler, "Memorandum for the Under Secretary: Proposed Presidential Statement on CB Warfare," Department of State, Deputy Undersecretary, 15 March 1967. "While we do not know how Defense will respond to [Walter] Rostow, in the past the Defense view has been that we should keep open the option for first use of incapacitating weapons (other than riot control agents). In the absence of a convincing case that the option is vital to national security, I feel that on balance, attempting to retain the option is not valid" (2).
37. "Letter from Secretary of Defense McNamara to the President's Special Assistant (Rostow)," 3 May 1967; Johnson Library, National Security File, Warfare, Chemical and Biological, 51. Copies were sent to Secretary of State Dean Rusk and to Foster at the Arms Control and Disarmament Agency.
38. Hersh, *Chemical and Biological Warfare,* 26–27.
39. Senator J. William Fulbright, head of the Foreign Relations Committee, sponsored crucial hearings on chemical and biological weapons throughout this period. Another critic, Senator Joseph Clark, became the first to use congressional power over the budget to restrain CWS funding. See McCarthy, *The Ultimate Folly,* 123–137, and Frank, "U.S. Arms Control Policymaking," 82.
40. See Philip Boffey and D. S. Greenberg, "6000 Sheep Stricken Near CBW Center," *Science* 159, no. 3822 (29 March 1968): 1442, for an early report.
41. Philip M. Boffey "Biological Warfare—Is the Smithsonian Really a Cover?" *Science* 163, no. 3869 (9 February 1969): 791.
42. Joshua Lederberg, "Congress Should Examine Biological Warfare Tests," *Washington Post,* 30 March 1968.
43. McCarthy, *The Ultimate Folly,* ix. Thanks to his aide, Wendell Pigman, McCarthy's book also reproduces documents from this time, such as the Eighteen Nations biological weapons treaty proposal.
44. Hersh, *Chemical and Biological Warfare,* 36. Among other journalists covering chemical and biological weapons issues were Phil Boffey, Anthony Lewis, Charles Mohr, Neil Sheehan, Robert M. Smith, and Nicholas Wade.
45. McCarthy noted that Venezuelan equine encephalitis virus was now found in Utah wildlife, far from its more typical Florida and Louisiana environments, and he questioned the environmental effects of open-air anthrax tests at Dugway and Eniewotok Atoll. McCarthy, *The Ultimate Folly,* 28–29.
46. Ibid., 102–108. McCarthy perhaps influenced his former colleague in the House of Representatives, Secretary of Defense Melvin Laird. On March 13, 1969, McCarthy personally insisted to Laird, with whom he had served in Congress, that he "wanted answers" about chemical and biological weapons (McCarthy, *The Ultimate Folly,* 140). On April 30, Laird requested that the NSC "immediately" issue a study of US chemical and biological warfare policy and programs (Laird, "Memorandum For: Assistant to the President for National Security Affairs," Secretary of Defense, 30 April 1969.) The NSC staff informed Kissinger that such a review was in process

(Morton H. Halperin, "Memorandum for Dr. Kissinger. Subject: Memorandum to Secretary Laird on CBW Study," 7 May 1969, White House).

47. Primack and von Hippel, "Matthew Meselson and Federal Policy," 152–154. Frederick Seitz, president of NAS and head of the Defense Department's top science advisory committee, the Defense Science Board, picked Kistiakowsky to chair the panel. Kistiakowsky insisted on picking the panel's members and included Meselson. As a member of the NAS panel, Meselson discovered by a visit to the Rocky Mountain Arsenal that technicians there had considerable experience in dismantling and destroying chemical stockpiles.
48. Henry A. Kissinger and Bernard Brodie, *Bureaucracy, Politics and Strategy* (Los Angeles: Security Studies Project, 1968), 9.
49. Frank, "U.S. Arms Control Policymaking," 78–79. Frank bases this movement on his interviews with anonymous government officials. According to Frank, prior to Nixon's taking office, there was also momentum in the previous administration to create a "Johnson peace legacy," which included the renunciation of biological weapons. If so, the opportunity to renounce biological warfare and renew treaty efforts effectively passed to Nixon.
50. Henry Kissinger, *The White House Years* (Boston: Little, Brown, 1979), 215–222. Kissinger describes Congress's trend toward military cutbacks and the Nixon administration's successes in financing the B-1 bomber, the Trident missile, and the cruise missile.
51. Walter Isaacson, *Kissinger: A Biography* (New York: Simon & Schuster, 1990), 204–205. Isaacson characterizes Nixon and Kissinger as two internationalists trying to scale back US responsibilities abroad to forestall full-blown isolationism in reaction to the Vietnam War.
52. McCarthy, *The Ultimate Folly*, 125.
53. "US Policy on Chemical and Biological Warfare and Agents," National Security Study Memorandum 59, National Security Council, 28 May 1969, 9.
54. Primack and von Hippel, "Matthew Meselson and Federal Policy," 150.
55. Matthew Meselson, "The United States and the Geneva Protocol of 1925," September 1969, author's personal files.
56. See Jonathan Tucker, "A Farewell to Germs: The U.S. Renunciation of Biological and Toxin Warfare, 1969–70," *International Security* 27, no. 1 (2002): 107–148.
57. Interdepartmental Political-Military Group, "US Policy on Chemical and Biological Warfare and Agents: Report to the National Security Council," 10 November 1969.
58. See Han Swyter, "Political Considerations and Analysis of Military Requirements for Chemical and Biological Weapons," *Proceedings of the National Academy of Sciences* 65, no. 1 (January 1970): 261–270.
59. Frank, "U.S. Arms Control Policymaking," 117–122. This distinction was later reflected in the wording of National Security Decision Memorandum 39 of 25 November 1969, announcing Nixon's decision to renounce biological weapons.
60. Nixon's message is reproduced in McCarthy, *The Ultimate Folly*, 169–171. McCarthy (143–144) recounts that this phrase about revulsion was deleted from the press re-

lease copy he first received and, after he protested, reinserted as it had been in the original.

61. Richard M. Nixon, "Remarks Announcing Decisions on Chemical and Biological Defense Policies and Programs," 25 November 1969, White House.
62. Hersh, *Chemical and Biological Warfare,* 38. In 1968, McNamara had attempted a policy review, but the effort foundered in disagreements between different State Department divisions; Frank, "U.S. Arms Control Policymaking," 64–65.
63. In the press conference after the November 25 announcement, where Kissinger spoke for the President, he handled all the questions about biological warfare expertly, stumbling only in his confusion about the definition of toxins. Later he would describe toxins as the issue that "fell between the bureaucratic cracks." See Frank, "U.S. Arms Control Policymaking," 280, and Tucker, "A Farewell to Germs," 136–138.
64. Matthew Meselson, "What Policy for Toxins," *Congressional Record,* 18 February 1979, E1042–E1043.
65. "British Working Paper on Microbiological Warfare," US Arms Control and Disarmament Agency, *Documents on Disarmament 1968* (Washington, DC: US Government Printing Office, 1969), 570.
66. Susan Wright, "Evolution of Biological Warfare Policy," in Susan Wright, ed., *Preventing a Biological Arms Race* (Cambridge, MA: MIT Press, 1990), 41.
67. For the BWC text, see Wright, *Preventing a Biological Arms Race,* Appendix C, 370–376.
68. Primack and von Hippel, "Matthew Meselson and Federal Policy," 160; Matthew S. Meselson, Arthur H. Westing, John D. Constable, and James E. Cook, "Preliminary Report of the Herbicide Assessment Commission," paper presented at the AAAS Annual Meetings, Chicago, 30 December 1970; reprinted in the *Congressional Record* 118 (1972): S3226–S3233.
69. Hearings Before the Committee on Foreign Relations, United States Senate, Ninety-Second Congress, First Session. "The Geneva Protocol of 1925" March 5, 16, 18, 19, 22, and 26 (Washington, DC: US Government Printing Office, 1972).
70. Ibid., "Statement of McGeorge Bundy, President, Ford Foundation," 183–194.
71. *U.S. Army Activity in the U.S. Biological Warfare Programs* 1 (24 February) (Washington, DC: Department of the Army, 1977), 227.
72. Frank, "U.S. Arms Control Policymaking," 285–290; Wright, "Evolution of Biological Warfare Policy," 42; see also Norman M. Covert, *Cutting Edge: A History of Fort Detrick, Maryland, 1943–1993* (Fort Detrick, MD: Public Affairs Office, 1993).
73. Hearings Before the Select Committee to Study Governmental Operations with Respect to Intelligence Activities of the United States Senate, Ninety-fourth Congress, First Session, vol. 1, "Unauthorized Storage of Toxic Agents" September 16, 17, and 18, 1975 (Washington, DC: US Government Printing Office, 1976).
74. See Raymond L. Garthoff, "Polyakov's Run," *Bulletin of the Atomic Scientists* 56, no. 5 (2000): 37–40.

7. The Soviet Biological Weapons Program

1. To sidestep the dual-use problem in the chemical industry, the framers of the 1925 Geneva Protocol opted for the ban on use. Later, at the League of Nations Disarmament Conference in 1932, the delegates advised a ban on all preparations for chemical warfare, requiring state militaries to report regularly all activities that might be relevant to either chemical or biological weapons. Then, in 1933, the United Kingdom proposed a comprehensive prohibition of chemical and biological weapons, banning retaliatory use and procedures for onsite investigation of violations. But, following its widely criticized invasion of Manchuria, Japan quit the League of Nations, and Germany, under Adolf Hitler's leadership, soon followed. In January 1936, the League of Nations postponed any further convocation of the Disarmament Conference, and it never met again. Stockholm International Peace Research Institute (SIPRI), *The Problem of Chemical and Biological Warfare*, vol. 5, *The Prevention of CBW* (New York: Humanities Press, 1971), 141–146.
2. Ibid., vol. 4, *CB Disarmament Negotiations, 1920–1970*, 147–159.
3. The Russian State Military Archive, Yale University, for example, is a major collection of Soviet documents.
4. See Valentin Bojtzov and Erhard Geissler, "Military Biology in the USSR, 1920–45," in Erhard Geissler and John Ellis van Courtland Moon, eds., *Biological and Toxin Weapons: Research, Development and Use from the Middle Ages to 1945* (New York: Oxford University Press, 1999), 153–167.
5. Col. Walter Hirsch, M.D., "Soviet BW and CW Preparations and Capabilities," US Army Chemical Intelligence Branch, 15 May 1951, Washington, DC (hereafter Hirsch Report).
6. Heinrich Kliewe, "Bacterial War" (translation of "Der Bakterienkrieg"), 19 January 1943, Alsos Mission, report No. C-H/303, War Department, Washington, DC, 1945.
7. On Pasechnik, see Simon Cooper, "Life in the Pursuit of Death," *Seed*, January/February 2003, 68–72, 104–107; Ken Alibek with Stephen Handelman, *Biohazard: The Chilling True Story of the Largest Covert Biological Weapons Program in the World* (New York: Random House, 1999); Igor V. Domaradskij and Wendy Orent, *Biowarrior: Inside the Soviet/Russian Biological War Machine* (Amherst, NY: Prometheus Press, 2003).
8. Hirsch Report, 84–85; John Buckley, *Air Power in the Age of Total War* (Bloomington: Indiana University Press, 1999), 100–101.
9. According to a Fishman progress report noted in Bojtzov and Geissler, "Military Biology in the USSR," 159.
10. Sally Stoecker, *Forging Stalin's Army: Marshal Tukhachevsky and the Politics of Military Innovation* (Boulder, CO: Westview Press, 1998), 91–93.
11. Hirsch Report, 101.
12. Ivan Velikanov, "Bacterial Warfare," in *Sovietskaja wojennaja enzilopedija* [Soviet Military Encyclopedia] (Moscow, 1933), 2:100–102; quoted in Bojtzov and Geissler, "Military Biology in the USSR," 155.

13. Hirsch Report, 84.
14. See David Jovarsky, *The Lysenko Affair* (Cambridge, MA: Harvard University Press, 1970); Valery N. Soyfer, *Lysenko and the Tragedy of Soviet Science,* trans. Leo Gruliow and Rebecca Gruliow (New Brunswick, NJ: Rutgers University Press, 1994).
15. Alibek, *Biohazard,* 37.
16. Domaradskij and Orent, "Memoirs," 248.
17. Anthony Rimmington, "The Soviet Union's Offensive Program: Implications for Contemporary Arms Control," in Susan Wright, ed., *Biological Weapons and Disarmament: New Problems* (New York: Rowman & Littlefield, 2001), 103–148. See also Alibek, *Biohazard,* 298–300. Production refers to growing the agents, not to loading munitions.
18. See Alibek, *Biohazard,* 300.
19. Ibid., 81.
20. Ibid., 234–235. Alibek centers on the US Army Medical Research Institute of Infectious Diseases (USAMRIID) as the source of Soviet suspicion, in addition to CIA defiance of Nixon's ban on biological weapons. He also links the US collusion with Japanese scientists directly to the building of the Sverdlovsk facility. See also Raymond L. Garthoff, "Polyakov's Run," *Bulletin of the Atomic Scientists* 56, no. 5 (2000): 37–40, for an analysis of US efforts to convince the Soviets of chemical and biological weapons threats.
21. See Alibek, *Biohazard,* 173.
22. Made before the Berlin Press Association, 13 September 1981, and published by the State Department as No. 311 in its series *Current Policy.* See also US Department of State Special Report No. 98, "Chemical Warfare in Southeast Asia and Afghanistan," 22 March 1982, and US Department of State Special Report No. 104, "Chemical Warfare in Southeast Asia and Afghanistan: An Update," 11 November 1982.
23. Meselson had long been interested in expert investigations and treaty compliance. See Seymour Hersh, *Chemical and Biological Warfare: America's Hidden Arsenal* (Indianapolis: Bobbs-Merrill, 1968), 308.
24. J. Nowicke and M. Meselson, "Yellow Rain: A Palynological Analysis," *Nature* 309, no. 5965 (1984): 205–206.
25. T. D. Seeley, J. W. Nowicke, M. Meselson, J. Guillemin, and P. Akratanakul, "Yellow Rain," *Scientific American* 253, no. 3 (1985): 128–137.
26. Julian Robinson, Jeanne Guillemin, and Matthew Meselson, "Yellow Rain in Southeast Asia: The Story Collapses," *Foreign Policy* 68 (fall 1987): 108–112.
27. Bangkok telegram 27244, US Embassy to Defense Intelligence Agency, Washington, 30 May 1984, subject: CBW Samples: TH-840523-IDS Through 7DS.
28. Bangkok telegram 11615, US Embassy to Defense Intelligence Agency, Washington, 6 March 1984, subject: CBW Sample TH-840209-1DL through TH-840209-11DL Supplemental Record.
29. Memorandum for record dated 31 August 1981 from SGMI-SA, subject: Telephone conversation with Fred Celec, State Department. See Robinson, Guillemin, and Meselson, "Yellow Rain," 227.

30. Soon after Haig's speech, media attention was given to two unofficial analyses. ABC-TV News had obtained a yellow rain sample and gave it to a Rutgers chemist, Joseph Rosen, for analysis. In December 1981, ABC announced a tricothecine finding, but its sample was later revealed as bee feces. See Nowicke and Meselson, "Yellow Rain," 205–206.
31. Robinson, Guillemin, and Meselson, "Yellow Rain in Southeast Asia," 110.
32. *House of Commons Official Report* 98, no. 117, col. 92, written answers to questions, 19 May 1986.
33. Alibek (*Biohazard,* 126–131) recounts the death of a Soviet scientist from an accidental laboratory infection of Marburg virus.
34. Dr. Pyotr Burgasov, interview, *60 Minutes,* CBS News, 11 May 2003. Dr. Burgasov was Deputy Minister of Health during the 1979 Sverdlovsk anthrax epidemic and a key figure in its containment. See Jeanne Guillemin, *Anthrax: The Investigation of a Deadly Outbreak* (Berkeley: University of California Press, 1999), 11–22. See also Richard D. McCarthy, *The Ultimate Folly* (New York: Vintage, 1969), 28–29, on US program accidents.
35. Throughout the 1980s Meselson put together expert teams to go to Sverdlovsk, only to have plans fail, due to Soviet reluctance (Guillemin, *Anthrax,* 92–93).
36. M. Meselson, J. Guillemin, M. Hugh-Jones, A. Langmuir, I. Popova, A. Shelokov, and O. Yampolskaya, "The Sverdlovsk Anthrax Outbreak of 1979," *Science* 266, no. 5188 (1994): 1202–1208; Guillemin, *Anthrax*; F. A. Abramova, L.M. Grinberg, O. V. Yampolskaya, and D. H. Walker, "Pathology of Inhalational Anthrax from the Sverdlovsk Outbreak in 1979," *Proceedings of the National Academy of Sciences* 90 (1993): 2291–2293.
37. Alibek, *Biohazard,* 80–81.
38. See, for example, Joshua Lederberg, ed., *Biological Weapons: Limiting the Threat* (Cambridge, MA: MIT Press, 1999), 31, 44, 78, 104.
39. Delayed diagnosis also influenced the 1972 smallpox outbreak in Yugoslavia, another model for possible bioterrorist attack, and the 2001 anthrax postal attacks. See Jeanne Guillemin, "Bioterrorism and the Hazards of Secrecy: A History of Three Epidemic Cases," *Harvard Health Policy Review* 1, no. 4 (2003): 36–50.
40. Guillemin, *Anthrax,* 163–166.
41. In 2000, General Stanislov Petrov, former head of Soviet biological and chemical troops, and others blamed CIA sabotage for the Sverdlovsk anthrax epidemic and accused the Meselson team of concocting its data. S. Petrov, M. Supotnitiskiy, and S. Vey, "Biological Diversion in the Urals," *Nezavisimaya Gazeta,* 23 May 2001, FBIS (Federal Broadcast Information Service) translation. As of December 1999, the Russian Deputy Minister of Health remained on record that the Sverdlovsk epidemic was caused by the distribution of anthrax-infected meat. G. G. Ochinnikov et al., *Siberskaya Yazva* (Moscow: Vunmic, 1999), 218–219, 236–237.
42. Guillemin, *Anthrax,* 163. In 2003 Biopreparat still maintained its Moscow headquarters and its commercial promotion of medical technologies, with some employment of former Fifteenth Directorate officials (author's field notes, June 4, 2003).

43. Stepnogorsk, in Kazakhstan, was not included in this declaration, nor were other resources outside the Russian Federation. See "Former Soviet Biological Weapons Facilities in Kazakhstan: Past, Present, and Future," CNS Occasional Papers, Monterey Institute.
44. This phrase was eliminated from the US State Department translation.
45. On accords, see Anthony H. Cordesman, *Terrorism, Asymmetric Warfare, and Weapons of Mass Destruction* (Westport, CT: Praeger, 2002), 205–206.
46. See the following two overviews. Amy Smithson, *Toxic Archipelago: Preventing Proliferation from the Former Soviet Chemical and Biological Weapons Complexes* (Washington, DC: Henry L. Stimson Center, 1999); and Anthony Rimmington, "From Offence to Defence? Russia's Reform of Its Military Microbiological Sector and the Implications for Western Security," *Journal of Slavic Military Studies* (March 2003): 78–98.
47. A.P. Pomerantsev, N.A. Staritsin, Y. V. Mockov, and L. I. Marinin, "Expression of Cereolysine AB Genes in *Bacillus anthracis* Vaccine Strain Ensures Protection Against Experimental Hemolytic Anthrax Infection," *Vaccine* 15 (December 1997): 1846–1850. The US media coverage of this research led to Pomerantsev's being temporarily restrained from returning to Russia from a trip to Japan; subsequently, by 2003, he was able to do research at the National Institutes of Health (personal communication).
48. Alibek, *Biohazard*, 121–122.
49. D. A. Henderson, "Pathogen Proliferation: Threats from the Former Soviet Bioweapons Complex," *Politics and the Life Sciences* 19, no. 1 (2000): 3–16.
50. Rimmington, "From Offence to Defence," 32. See Judith Miller, Stephen Engelberg, and William Broad, *Germs: Biological Weapons and America's Secret War* (New York: Simon & Schuster, 2001), 165–168.
51. David Kelly, "The Trilateral Agreement: Lessons for Biological Weapons Verification" in Trevor Findlay and Oliver Meier, eds., *The Verification Yearbook, 2002* (London: VERTIC, 2002), 98–102.
52. See Benoit Morel and Kyle Olson, eds., *Shadows and Substance: The Chemical Weapons Convention* (Boulder, CO: Westview), 1993; *Ad Hoc Group of Governmental Experts to Identify and Examine Potential Verification Measures from a Scientific and Technical Standpoint Report*, Geneva, 1993.
53. Oliver Thränert, "The Compliance Protocol and the Three Depository Powers," in Wright, *Biological Weapons and Disarmament*, 343–368.

8. Bioterrorism and the Threat of Proliferation

1. *Ad Hoc Group of Governmental Experts to Identify and Examine Potential Verification Measures from a Scientific and Technical Standpoint Report*, Geneva, 1993. See also Marie Isabelle Chevrier, "Preventing Biological Proliferation: Strengthening the Biological Weapons Convention," in Oliver Thränert, ed., *Preventing the Proliferation of Weapons of Mass Destruction: What Role for Arms Control?* (Berlin: Friedrich-Ebert-Stiftung, 1999), 85–98; Jonathan B. Tucker, "Strengthening the BWC:

Moving Toward a Compliance Protocol," *Arms Control Today*, January/February 1998, 20–27.

2. See Richard A. Falkenrath, Robert D. Newman, and Bradley A. Thayer, *America's Achilles' Heel: Nuclear, Biological, and Chemical Terrorism and Covert Attack* (Cambridge, MA: MIT Press, 1998); Anthony H. Cordesman, *Terrorism, Asymmetric Warfare, and Weapons of Mass Destruction* (Westport, CT: Praeger, 2002); Laura Drake, "Integrated Regional Approaches to Arms Control and Disarmament," in Susan Wright, ed., *Biological Warfare and Disarmament: New Problems/New Perspectives* (New York: Rowman & Littlefield, 2003), 151–180. As a general description of nuclear, chemical, and biological weapons, the term "weapons of mass destruction" requires explanation. Since it refers to weapons that can be or in the case of nuclear and biological weapons development have been aimed indiscriminately at civilians, it is used here and throughout the following chapters.
3. On modern "totalitarian" terrorism, see Michael Walzer, *Just and Unjust War: A Moral Argument with Historical Illustrations* (New York: Basic Books, 1980), 197–204.
4. Ulrich Beck, *Risk Society. Towards a New Modernity* (London: Sage, 1992), 76–78. See also Nicholas Benjamin King, "The Influence of Anxiety: September 11, Bioterrorism, and American Public Health," *Journal of the History of Medicine and Allied Sciences* 58, no. 4 (2003): 433–441.
5. Spencer R. Weart, *Nuclear Fear: A History of Images* (Cambridge, MA: Harvard University Press, 1988), 254–256. Weart describes the shelter movement collapsing within a year as the United States, United Kingdom, and Soviet Union moved to a new spirit of cooperation in 1963, following the Cuban missile crisis. He points out that the agreement not to test nuclear weapons in the atmosphere calmed the public but did not stop the arms race (259–260).
6. See Graham T. Allison and Philip Zelikow, *Explaining the Cuban Missile Crisis*, 2d ed. (Glenview, IL: Longman, 1999).
7. Rebecca S. Bjork, *The Strategic Defense Initiative: Symbolic Containment of the Nuclear Threat* (Albany: State University of New York Press, 1992), 52.
8. Ibid., 74–76. According to Bjork, SDI was considered as protection against threats from small states with long-range missile capacity (such as Iraq), especially by then Defense Secretary Richard Cheney. After the Gulf War, President George H. W. Bush redirected SDI toward providing protection from limited ballistic missile strikes (106–107). The public was not advised that SDI would give the US defensive cover for first or second nuclear strikes, and Reagan himself seemed unclear about the obviously destabilizing effect this system could have on world politics.
9. Brian Balmer, *Britain and Biological Warfare: Expert Advice and Science Policy, 1930–65* (London: Palgrave, 2001), 72–73. See Barry S. Levy and Victor W. Sidel, *Terrorism and Public Health. A Balanced Approach to Strengthening Systems and Protecting People* (New York: Oxford University Press, 2003), 3–18.
10. Stuart E. Johnson, "Introduction," in Stuart E. Johnson, *The Niche Threat: Deterring the Use of Chemical and Biological Weapons* (Washington, DC: National Defense University Press, 1977), 3. In the same volume, see Brad Roberts, "Between Panic

and Complacency: Calibrating the Chemical and Biological Warfare Problem," 9–42.

11. A comprehensive list of suspect countries, with bibliography, is contained in *Chemical and Biological Weapons: Possession and Programs Past and Present* (Monterey, CA: Monterey Institute of International Studies, 2002).
12. This overview is taken from Avner Cohen, "Israel and Chemical/Biological Weapons: History, Deterrence, and Arms Control," *The Nonproliferation Review* fall/winter 2001, 27–53. Information about Israel's biological and chemical weapons activities, as with its nuclear weapons, has remained highly secret. The 1998 crash of an El Al plane near Amsterdam let to the discovery that a dual-use chemical for making sarin nerve gas had been on board. Israel was legally pressured to reveal the full nature of twenty tons of cargo, which burned after the crash and caused unusual sickness in the local community. Israel's refusal to comply brought attention to its chemical and biological programs, since the shipment appears to have been destined for IIBR. In 1998, the Institute became the focus of controversy when residents of Ness Ziona protested the center's planned expansion as an environmental hazard. The public protest, which halted the project, did not extend to the chemical and biological weapons program itself.
13. Michael Klare, *Rogue States and Nuclear Outlaws* (New York: Hill and Wang, 1995), 26–28.
14. Graham S. Pearson, *The UNSCOM Saga: Chemical and Biological Weapons Non-Proliferation* (New York: St. Martin's Press, 1999), 11. See also Raymond S. Zilinskas, "Iraq's Biological Weapons Program: The Past as Future?" *Journal of the American Medical Association* 278, no. 5 (1997): 418–424.
15. Stephen Black, "Investigating Iraq's Biological Weapons Program," in Joshua Lederberg, ed., *Biological Weapons: Limiting the Threat* (Cambridge, MA: MIT Press, 1999), 159–163.
16. Ibid.
17. United Nations, "UNSCOM/IAEA Interview with General Kamal," 22 August 1995, 7. See also Hans Blix, *Disarming Iraq* (New York: Pantheon, 2004), 29–30.
18. Richard Butler, *The Greatest Threat: Iraq, Weapons of Mass Destruction, and the Crisis of Global Security* (New York: Public Affairs, 2000).
19. Scott Ritter, *Endgame: Solving the Iraq Problem—Once and for All* (New York: Simon & Schuster, 1999). Butler addresses Ritter's resignation in *The Greatest Threat,* 178–185.
20. United Nations, *UNSCOM's Comprehensive Review: Status of Verification of Iraq's Biological Warfare Programme* (New York: United Nations, 1998), annex C, item 15.
21. Two senior UNSCOM participants were David Kelly from the United Kingdom and Nikita Smidovich from Russia, who had also been part of the tripartite exchange.
22. United Nations, *UNSCOM's Comprehensive Review,* item 175.
23. Butler, *The Greatest Threat,* 157–159.
24. Chandré Gould and Peter Folb, *Project Coast: Apartheid's Chemical and Biological Warfare Program* (Geneva: United Nations Institute for Disarmament Research, 2002), 11.

25. Ian Martinez, "The History of the Use of Bacteriological and Chemical Agents during Zimbabwe's Liberation War of 1965–80 by Rhodesian Forces," *Third World Quarterly* 23, no. 6 (2002): 1159–1179.
26. Chandré Gould and Peter Folb, "The Rollback of the South African Biological Weapons Program," *INSS Occasional Paper 37*, February 2001, 1–114; Gould and Folb, *Project Coast*, 203. Basson also hired Larry Ford, an American physician who was a right-wing extremist and collaborated on medical projects with Basson. Ford committed suicide at his home in California in 2000. After his death, authorities discovered a small arsenal of conventional weapons, a large container of potassium cyanide, and stores of three pathogens, for cholera, typhoid fever, and brucellosis.
27. See Gould and Folb, *Project Coast*, 223–230.
28. See Paul Farmer, *Pathologies of Power: Health, Human Rights, and the New War on the Poor* (Berkeley: University of California Press, 2002).
29. John V. Parachini, "The World Trade Center Bombers (1993)," in Jonathan Tucker, ed., *Toxic Terror: Assessing Terrorist Use of Chemical and Biological Weapons* (Cambridge, MA: MIT Press, 2000), 185–226.
30. Cordesman, *Terrorism, Asymmetric Warfare, and Weapons of Mass Destruction*, 7, 247–249. Terrorist loners, like McVeigh and the so-called Unabomber, Theodore Kaczynski, alerted the FBI to individuals who might be bioterrorists.
31. Daniel Benjamin and Steven Simon, *The Age of Sacred Terror* (New York: Random House, 2003), 230.
32. Jonathan B. Tucker and Jason Pate, "The Minnesota Patriots Council (1991)," in Jonathan B. Tucker, ed., *Toxic Terror: Assessing Terrorist Use of Chemical and Biological Weapons* (Cambridge, MA: MIT Press, 2002), 159–184.
33. Jessica Eve Stern, "Larry Wayne Harris (1998)," in Tucker, *Toxic Terror*, 227–246.
34. See David E. Kaplan and Andrew Marshall, *The Cult at the End of the World* (New York: Crown, 1996); Robert Jay Lifton, *Destroying the World in Order to Save It* (New York: Metropolitan Books, 1999).
35. Seth Carus, "The Rajneeshees (1984)," in Tucker, *Toxic Terror*, 115–137.
36. The US Senate held hearings on the Aum Shinrikyo, the testimony for which was contested by Milton Leitenberg. See *1995 Hearings on Global Proliferation of Weapons of Mass Destruction: A Case Study of the Aum Shinrikyo, Oct. 31* (Washington, DC: US Government Printing Office, 1995), 41–44, and *US Senate 1996 Committee on Government Affairs Global Proliferation of Weapons of Mass Destruction, Part I* (Washington, DC: US Government Printing Office, 1996), 273–276.
37. Testimony, February 4, 1999. Senate Committee on Appropriations Subcommittee for the Departments of Commerce, Justice, and State, the Judiciary, and Related Agencies.
38. See the balanced analysis in Mark Juergensmeyer, *Terror in the Mind of God: The Global Rise of Religious Violence* (Berkeley: University of California Press, 2000).
39. Quoted in Gideon Rose, "It Could Happen Here: Facing New Terrorism" *Foreign Affairs* 78, no. 2 (1999): 131–137.
40. Walter Laqueur, *The New Terrorism: Fanaticism and the Arms of Mass Destruction* (New York: Oxford University Press, 1999).

41. Samuel P. Huntington, *The Clash of Civilizations and the Remaking of World Order* (New York: Simon & Schuster, 1996).
42. Benjamin and Simon, *The Age of Sacred Terror,* 256–262.
43. Ibid., 357–359; Butler, *The Greatest Threat,* 212–213.
44. On preemptive war, accidental causes, and evidence, see Stephen Van Evera *Causes of War: Power and the Roots of Conflict* (Ithaca, NY: Cornell University Press, 1999), 42–43.
45. Benjamin and Simon, *The Age of Sacred Terror,* 236–239, 265; Judith Miller, Stephen Engelberg, and William Broad, *Germs: Biological Weapons and America's Secret War* (New York: Simon & Schuster, 2001), 231–233; Clarke, *Against All Enemies,* 198–204.
46. Richard Danzig, "Biological Warfare: A Nation at Risk—A Time to Act," *INSS Strategic Forum 58* (Washington, DC: National Defense University, January 1996); Richard Danzig and Pamela B. Berkowsky, "Why Should We Be Concerned About Biological Warfare?" *Journal of the American Medical Association* 278, no. 5 (1997): 431–432.
47. Lederberg, a government insider, did not sign the 1966 Meselson-Edsall petition and was ambivalent about the US ratification of the Geneva Protocol as it might limit military options to use chemicals in war. See Joshua Lederberg, "Letter and Enclosed Statement to Senator J. W. Fulbright," *Hearings on Geneva Protocol of 1925, Committee on Foreign Relations, US Senate* (Washington, DC: US Printing Office, 1972), 425–427. See also Miller, Engelberg, and Broad, *Germs,* 140.
48. Joshua Lederberg, "Mankind Had a Near Miss from a Mystery Pandemic," *Washington Post,* 7 September 1968.
49. Quoted in C. J. Peters and Mark Olshaker, *Virus Hunter: Thirty Years of Battling Hot Viruses Around the World* (New York: Anchor Books, 1997), 298.
50. Miller, Engelberg, and Broad, *Germs,* 223–234.
51. Ibid., 198–199.
52. Ibid., 198.
53. Leonard A. Cole, "Risk of Publicity about Bioterrorism: Anthrax Hoaxes and Hype," *American Journal of Infection Control* 27 (December 1999): 470–473. Some of the aggression was free-floating, against co-workers or estranged partners; some was directed against Jews and abortion clinics. A handful involved boys who wanted to get out of school for a day. The highly publicized arrest in February 1998 of Larry Wayne Harris helped spur the hoaxes as well.
54. Editorial, "Thwarting Tomorrow's Terrors," *New York Times,* 23 January 1999. Despite the Act, the US military has a spotty history of being deployed to quell civil unrest, as in 1932 when it stopped the protest of World War I veterans for benefits and in 1943 when it put down black riots in Detroit. In 1968, after Martin Luther King was assassinated, 21,000 soldiers were dispersed to American cities to maintain civil order, and in 1992 Marines were sent to Los Angeles to protect police in the Rodney King riots.
55. Amy E. Smithson and Leslie-Anne Levy, *Ataxia: The Chemical and Biological Terrorism Threat and the US Response* (Washington, DC: Henry L. Stimson Center, 2000), 154.

56. Tara O'Toole and Thomas Inglesby of the Johns Hopkins Center for Civilian Biodefense Studies and Randy Larsen and Mark DeMier of Analytic Services (ANSER) are listed as principal designers. The Center for Strategic and International Studies and the Memorial Institute for the Prevention of Terrorism were cosponsoring organizations. See Tara O'Toole, Michael Mair, and Thomas V. Inglesby, "Shining Light on 'Dark Winter,'" *Clinical Infectious Diseases* 34 (2002): 972–983.
57. Martin I. Meltzer, Inger Damon, James W. LeDuc, and J. Donald Millar, "Modeling Potential Responses to Smallpox as a Bioterrorist Weapon," *Emerging Infectious Diseases* 7, no. 6 (2001): 959–969; Thomas Mack, "A Different View of Smallpox and Vaccination," *New England Journal of Medicine* 348, no. 5 (2003), 1–4. For a critique from Porton, see Raymond Ganl and Steve Leach, "Transmission Potential of Smallpox in Contemporary Populations," *Nature* 414, no. 13 (2001): 748–751.
58. "FEMA's Role in Managing Bioterrorism Attacks and the Impact of Public Health Concerns on Bioterrorism Preparedness," U.S. Senate Government Affairs Subcommittee Hearing on International Security, Proliferation and Federal Services, 23 July 2001. The Bush administration proved more supportive of vaccine production and stockpiling than the Clinton administration. See Cordesman, *Terrorism, Asymmetric Warfare, and Weapons of Mass Destruction,* 252–269.
59. Robert J. Blendon, Catherine M. Des Roches, John M. Benson, Melissa J. Hermann, et al., "The Public and the Smallpox Threat," *New England Journal of Medicine* 348, no. 5 (2002): 426–432.
60. David Koplow, *Smallpox: The Fight to Eradicate a Global Scourge* (Berkeley: University of California Press, 2003), 193–204; Jonathan Tucker, *Scourge: The Once and Future Threat of Smallpox* (New York: Atlantic Monthly Press, 2001), 190–230.
61. Tucker, *Scourge,* 182.
62. Ibid., 201.
63. National Institute of Medicine, *Assessment of Future Scientific Needs for Live Variola Virus* (Washington, DC: National Academy Press, 1999).
64. John Duffy, *The Sanitarians: A History of American Public Health* (Urbana: University of Illinois Press, 1990). Noting that the United States is well below other countries in measures of public health, Duffy observes, "This fact can be accounted for in part by America's firm commitment to rugged individualism and personal liberty and to the general suspicion of government controls." (313–314). See also Laurie Garrett, *Betrayal of Trust: The Collapse of Global Public Health* (New York: Hyperion, 2000).
65. Donna Shalala, head of HHS, stated, "This is the first time in American history in which the public health system has been directly integrated into the national security system." US White House Office of the Press Secretary transcript of press briefing, 22 January 1999, with Janet Reno, Attorney General, Donna Shalala, Secretary of Health and Human Services, and Richard Clarke, President's National Coordinator for Security, Infrastructure, and Counterterrorism.
66. Victor W. Sidel, "Defense Against Biological Weapons: Can Immunization and Secondary Prevention Succeed?" in Wright, *Biological Warfare and Disarmament,* 77–101. See also Barry S. Levy and Victor W. Sidel, "Challenges that Terrorism Poses to

Public Health," in *Terrorism and Public Health: A Balanced Approach to Strengthening Systems and Protecting People* (New York: Oxford University Press, 2003), 3–18.

67. Jeanne Guillemin, "Soldiers' Rights and Medical Risks: The Protest Against Universal Anthrax Vaccinations," *Human Rights Review* 1, no. 4 (2000): 124–139; and "Medical Risks and the Volunteer Army," in Pamela R. Frese and Margaret C. Harrell, *Anthropology and the United States Military: Coming of Age in the Twenty-first Century* (New York: Palgrave, 2003), 29–44.
68. Jeanne Guillemin, "Bioterrorism and the Hazards of Secrecy: A History of Three Epidemic Cases," *Harvard Health Policy Review* 4, no. 1 (2003): 36–50.

9. National Security and the Biological Weapons Threat

1. Judith Miller, Stephen Engelberg, and William Broad, *Germs: Biological Weapons and America's Secret War* (New York: Simon & Schuster, 2001), 293.
2. A.P. Pomerantsev, N.A. Staritsin, Y. V. Mockov, and L. I. Marinin, "Expression of Cereolysine AB Genes in *Bacillus anthracis* Vaccine Strain Ensures Protection against Against Experimental Hemolytic Anthrax Infection," *Vaccine* 15 (December 1997): 1846–1850.
3. Miller, Engelberg, and Broad, *Germs,* 310.
4. Susan Wright and Richard Falk, "Rethinking Biological Disarmament," in Susan Wright, ed., *Biological Warfare and Disarmament: New Problems/New Perspectives* (New York: Rowman & Littlefield, 2002), 413–440.
5. John Steinbrunner, "Confusing Ends and Means: The Doctrine of Coercive Preemption," *Arms Control Today* 33, no. 1 (2003): 3–5. Hans Blix, head of UNMOVIC, agreed and gave his perspective in detail in *Disarming Iraq* (New York: Pantheon), 2004.
6. Bob Woodward, *Bush at War* (New York: Simon & Schuster, 2003), 48–49; Richard A. Clarke, *Against All Enemies: Inside America's War on Terror* (New York: Free Press, 2004), 30–33, 264–273.
7. See John Cassidy, "Letter from London: The David Kelly Affair," *New Yorker,* 8 December 2003.
8. Seth Carus, "Prevention Through Counter-Proliferation," in Ray Zilinskas, ed., *Biological Warfare: Modern Offense and Defense* (Boulder, CO: Lynne Rienner, 1999), 194–196.
9. In his speech Blix reviewed the history, including the December 1999 UN resolution 1284 that asked for cooperation in return for lifting sanctions, which Iraq ignored. Associated Press, "Text of the U.N. Monitoring, Verification and Inspection Commission Executive Chairman Hans Blix's statement to the United Nations on Monday on weapons inspections in Iraq," 27 January 2003. For his fuller account, see Blix, *Disarming Iraq,* 145–150.
10. Blix, *Disarming Iraq,* 151–178.
11. Hearing of the Senate Armed Services Committee: Iraqi Weapons of Mass Destruction, January 28, 2004. David Kay testified, regarding Saddam's arsenal, "We were almost all wrong."

12. On the Bush speech, see Christopher Marquis, "How Powerful Can 16 Words Be?" *New York Times,* 20 July 2003.
13. Amy E. Smithson and Leslie-Anne Levy, *Ataxia: The Chemical and Biological Terrorism Threat and the US Response* (Washington DC: Henry L. Stimson Center, 2000); General Accounting Office, *Bioterrorism: Federal Research and Preparedness Activities* GAO-01-915 (Washington, DC: General Accounting Office) September 2001; Richard A. Falkenrath, "Problems of Preparedness: U.S. Readiness for a Domestic Terrorist Attack," *International Security* 25, no. 4 (spring 2001): 147–186; *Report of the Commission to Assess the Organization of the Federal Government to Combat the Proliferation of Weapons of Mass Destruction* (Washington, DC: US Government Printing Office, 1999). Falkenrath was on the White House staff as a contributor to the framing of the Homeland Security Act of 2003.
14. Information on the anthrax postal attacks is based on the author's field and literature research October 2001 through July 2003. See Jeanne Guillemin, "Bioterrorism and the Hazards of Secrecy: A History of Three Epidemic Cases," *Harvard Health Policy Review* 4, no. 1 (2003): 36–50. An overview of the 2001 anthrax attacks by the author is in the 2004 World Health Organization publication *Public Health Response to Biological and Chemical Weapons: WHO Guidance*, appendix 4.3 (98–108) "The Deliberate Release of Anthrax Spores though the United States Postal System." See also Leonard A. Cole, *The Anthrax Letters: A Medical Detective Story* (Washington, DC: National Academies Press, 2003); Marilyn W. Thompson, *The Killer Strain: Anthrax and a Government Exposed* (New York: HarperCollins, 2003).
15. B. Kournikakis, S. J. Armour, C. A. Boulet, M. Spence, and B. Parsons, "Risk Assessment of Anthrax Threat Letters Defence," Research Establishment Suffield (DRES TR-2001-048), September 2001.
16. Scott Shane, "Md. Experts' Key Lessons on Anthrax Go Untapped," *Baltimore Sun,* 4 November 2001.
17. Matthew Meselson, "Note Regarding Source Strength" *ASA Newsletter,* 21 December 2001, 10–11.
18. Daniel B. Jernigan, Pratima L. Raghunathan, Beth P. Bell, Ross Brechner, et al., "Investigation of Bioterrorism-Related Anthrax, United States. Epidemiologic Findings," *Emerging and Infectious Disease* 8, no. 10 (2002): 1019–1028.
19. Philip S. Brachman, "The Public Health Response to the Anthrax Epidemic," in Barry S. Levy and Victor W. Sidel, eds., *Terrorism and Public Health: A Balanced Approach to Strengthening Systems and Protecting People* (New York: Oxford University Press, 2003), 101–117. See Diane Vaughan's analysis of the "normalization of deviance," *The Challenger Launch Decision: Risky Technology, Culture, and Deviance at NASA* (Chicago: University of Chicago Press, 1997); and James R. Chiles, *Inviting Disaster: Lessons from the Edge of Technology* (New York: HarperCollins, 2002).
20. US Department of Homeland Security, *Homeland Security Exercise and Evaluation Program,* vol. I, *Overview and Doctrine* (Washington, DC: Department of Homeland Security, Office of Domestic Preparedness, 2003).
21. Ibid., appendix I, 5.

22. Advisory Panel to Assess Domestic Response Capabilities for Terrorism Involving Weapons of Mass Destruction, *Fourth Annual Report, Implementing the National Strategy* (Washington, DC: National Defense Research Institute, 2002), appendix P, 1–3.
23. For an overview of federal agencies tasked to WMD antiterrorism, see Anthony H. Cordesman, *Terrorism, Asymmetric Warfare, and Weapons of Mass Destruction* (Westport, CT: Praeger, 2002), 275–372.
24. Benjamin Riley, "Information Sharing in Homeland Security and Homeland Defense: How the Department of Defense Is Helping," *Journal of Homeland Security*, September 2003, 6. New exercises with NORTHCOM are promised that will be larger and more geographically dispersed than past demonstrations.
25. See James G. Hodge Jr. and Lawrence O. Gostin, "Protecting the Public's Health in an Era of Bioterrorism," in Jonathan A. Moreno, ed., *In the Wake of Terror: Medicine and Morality in a Time of Crisis* (Cambridge, MA: MIT Press, 2003), 17–32.
26. For example, Anthony Fauci argued that preparations against bioterrorism should include "classic public health activities at the federal and local levels" including hospital and public health facilities. "Foreword," in Donald A. Henderson, Thomas V. Inglesby, and Tara O'Toole, eds., *Bioterrorism: Guidelines for Medical and Public Health Management* (Chicago: AMA Press, 2002), vii–viii.
27. Richard Pilch, "Smallpox: The Disease versus the Vaccine." Center for Non-Proliferation Studies, Monterey Institute of International Studies, 3 February 2003.
28. Emily Martin, *Flexible Bodies: The Role of Immunity in American Culture from the Days of Polio to the Age of AIDS* (Boston: Beacon Press, 1994), 196–203; see also Mary Douglas on the immune system and society, "The Self as Risk-taker," in *Risk and Blame: Essays in Cultural Theory* (London: Routledge, 1992), 102–121.
29. Richard V. Neustadt and Harvey V. Feinberg, *The Swine Flu Affair: Decision Making on a Slippery Slope* (Washington, DC: US Department of Health, Education, and Welfare, 1978); Arthur M. Silverstein, *Pure and Impure Science: The Swine Flu Affair* (Baltimore: Johns Hopkins University Press, 1981).
30. Victor Sidel, Meryl Nass, and Todd Ensign, "The Anthrax Dilemma," *Medicine and Global Security* 2, no. 5 (1998): 97–104. See also Institute of Medicine, *Anthrax Vaccine: Is It Safe? Does It Work?* (Washington, DC: National Academy Press, 2002), 92–105. Unlike the live vaccines used in Russia and China, AVA (anthrax vaccine adsorbed) is acellular. By subcutaneous injection, it introduces into the body one of the three anthrax toxin proteins, PA (protective antigen), which, with aluminum hydroxide, stimulates antibodies. Six inoculations over an eighteen-month period seem to produce full immunity; even three shots might convey nearly complete resistance, although tests of human inhalational anthrax are lacking. As critics have pointed out, the vaccine's long-term effects were never studied; relatively few people had taken it. In June 2002 the Bush administration announced the stockpiling of AVA for civilian use. See also Arthur Friedlander, S. L. Welkos, M. L. Pitt, J. W. Ezzell, et al., "Postexposure Prophylaxis against Experimental Inhalational Anthrax," *Journal of Infectious Diseases* 167, no. 5 (1993): 691–702.
31. C. Dixon, *Smallpox* (London: J & A Churchill, 1962), 1460.

32. Dryvax was manufactured in the 1980s from calf lymph containing live *vaccinia* virus, much less virulent than smallpox virus. It contained four antibiotics (polymeric B, streptomycin, tetracycline, and neomycin) and was diluted with glycerin and phenol. If in limited supply, the vaccine could be diluted and still offer protection.
33. For an early report of this possible side effect, see J. B. Dalgaard, "Fatal Myocarditis Following Smallpox Virus," *American Heart Journal* 54 (1957): 156–157.
34. John Bartlett, Luciana Borio, Lew Radonovich, Julie Samia Mair, et al., "Smallpox Vaccination in 2003: Key Information for Clinicians," *Clinical Infectious Diseases* 36 (2003): 883–902. This article covers legal liability issues, which are protective of those who administer the vaccine, and outlines the recent Israeli smallpox vaccination campaign in which fifteen thousand people, almost all healthcare workers, were vaccinated. The lead authors and the journal section editors for this article are from the Johns Hopkins Center for Civilian Biodefense Strategies, which later moved to University of Pittsburgh Medical Center and became the Center for Biosecurity.
35. Public Health Security and Bioterrorism Preparedness and Response Act of 2002, Public Law 107–188.
36. Anthony Fauci, "An Expanded Biodefense Role for the National Institutes of Health," *Journal of Homeland Security* (April 2002): 1–3.
37. The Patriot Act had delegated the control of select biological agents to the Attorney General, with HHS as an advisor.
38. Editorial, "The End of Innocence?" *Nature* 414 (15 November 2001): 236.

10. Biological Weapons

1. Graham S. Pearson, "Biological Weapons: The British View," in Brad Roberts, ed., *Biological Weapons: Weapons of the Future?* (Washington, DC: CSIS, 1993), 7–18. See also Malcolm Dando, *The New Biological Weapons: Threat, Proliferation, and Control* (Boulder, CO: Lynne Rienner, 2002); and Mark Wheelis and Malcolm Dando, "Back to Bioweapons?" *Bulletin of the Atomic Scientists* 59 (2003): 40–46.
2. See Jessica Stern, "Dreaded Risks and the Control of Biological Weapons," *International Security* 27, no. 3 (2003): 89–123.
3. Bob Woodward, *Bush at War* (New York, Simon & Schuster, 2003), 293.
4. Susan Wright, "Introduction," in Susan Wright, ed., *Biological Warfare and Disarmament: New Problems/New Perspectives* (New York: Rowman & Littlefield, 2002), 3–24.
5. Daniel Feakes, "Global Civil Society and Biological and Chemical Weapons," in Helmut Anheier, Marlies Glasius, and Mary Kaldor, eds., *Global Civil Society Yearbook 2003* (Oxford: Oxford University Press, 2003), 87–117.
6. Douglas J. MacEachin, "Routine and Challenge: Two Pillars of Verification," *The CBW Conventions Bulletin* 39 (March 1998): 1–3.
7. Susan Wright and David Wallace, "Secrecy in the Biotechnology Industry," in Wright, *Biological Warfare and Disarmament*, 369–390.

8. See Sheila Jasanoff, "Three Cultures and the Regulation of Technology," in Martin Bauer, ed., *Resistance to New Technology: Nuclear Power, Information Technology, and Biotechnology* (Cambridge: Cambridge University Press, 1995), 311–334.
9. Associated Press, "Pharmaceutical Companies Sue South Africa over Patent Law" 5 March 2001. The suit was later dropped and an import agreement reached. Within a month, five major companies announced a cooperative project with the WHO: "United Nations Secretary-General to Lead Fight Against AIDS," Press Release SG/2070, AIDS4, 4 April 2001.
10. See Ulrich Beck, *Risk Society: Towards a New Modernity* (London: Sage, 1992): 76. From this German perspective, the public gets a say in technical matters and businesses may find themselves "on the witness bench" or "locked in the pillory."
11. Not all biotechnology leaders agreed. A former US Army Medical Research Institute of Infectious Diseases (USAMRIID) physician who started his own biotechnology company wrote to *Science* in support of the BWC compliance measures. See Thomas Monath and Lance Gordon, "Strengthening the Biological Weapons Convention," *Science* 282 (20 November 1998): 1423. A conference at the Stimson Center in Washington, DC, showed that individual executives were more amenable than PhRMA's position suggested. Stimson Center, *Compliance Through Science: US Pharmaceutical Industry Experts on a Strengthened Bioweapons Nonproliferation Regime* (Washington, DC: Stimson Center, 2002).
12. On chemical industry support, see the editorial, "The CWC and the BWC. Yesterday, Today, and Tomorrow," *CBW Conventions Bulletin* 50 (December 2000): 1–2.
13. "News Chronology," *The CBW Conventions Bulletin* 57 (September 2002): 47.
14. Nicholas A. Sims, "A Proposal for Putting the 26 March 2005 Anniversary to Best Use for the BWC," *The CBW Conventions Bulletin* 62 (December 2003): 1–6.
15. Angela Woodward, *Time to Lay Down the Law: National Legislation to Enforce the BWC* (London: VERTIC, 2003), 13. Article IV reads, "Each State Party to this Convention shall, in accordance to its constitutional processes, take any necessary measures to prohibit and prevent the development, production, stockpiling, acquisition or retention of the agents, toxins, weapons, equipment and means of delivery specified in article I of the Convention, within the territory of such State, under its jurisdiction or under its control anywhere." In Wright, *Biological Warfare and Disarmament*, 372.
16. Matthew Meselson and Julian Perry Robinson, "Draft Convention to Prohibit Biological and Chemical Weapons Under International Criminal Law," in R. Yepes-Enríquez and L. Tabassi, eds., *Treaty Enforcement and International Cooperation in Criminal Matters* (The Hague: OPCW, 2002), 457–469.
17. Rome Statute of the International Criminal Court, Article 8. A/CONF.183/9*, 17 July 1998.
18. See Julian Perry Robinson, "Chemical and Biological Weapons Proliferation and Control," in Elizabeth Clegg, Paul Eavis, and John Thurlow, eds., *Proliferation and Export Controls: An Analysis of Sensitive Technologies and Countries of Concern* (London: Deltac Limited and Saferworld, 1995), 29–53.

19. US Department of State, "Proliferation Security Initiative: Statement of Interdiction Principles," 4 September 2003.
20. John Bolton, Under Secretary of State, "Proliferation Brief," 2 December 2003, Institute for Foreign Policy Analysis, Washington, DC. Available with other information on arms control at the website of the Carnegie Endowment for International Peace, http://www.ceip.org.
21. John D. Steinbrunner and Elisa D. Harris, "Controlling Dangerous Pathogens," *Issues in Science and Technology* (spring 2003): 47–54; see also Jonathan Tucker, "Regulating Scientific Research of Potential Relevance to Biological Warfare," in Michael Barletta, ed., *After 9/11: Preventing Mass-Destruction Terrorism and Weapons Proliferation* (Monterey, CA: Monterey Center for Nonproliferation Studies, 2002), 24–27; and Claire M. Fraser and Malcolm R. Dando, "Genomics and Future Biological Weapons: The Need for Preventative Action by the Biomedical Community," *Nature Genetics* 29 (2001): 253–265.
22. Oliver Thränert, "The Compliance Protocol and the Three Depository Powers," in Wright, *Biological Weapons and Disarmament*, 343–368.
23. Monica Schoch-Spana, "Educating, Informing, and Mobilizing the Public," in Barry S. Levy and Victor W. Sidel, eds., *Terrorism and Public Health: A Balanced Approach to Strengthening Systems and Protecting People* (New York: Oxford University Press, 2003), 118–135.
24. Thomas Schelling, "Foreword," in Roberta Wohlstetter, *Pearl Harbor: Warning and Decision* (Stanford, CA: Stanford University Press, 1967), viii.
25. Martin S. Smolinski, Margaret A. Hamburg, and Joshua Lederberg, eds., *Microbial Threats to Health: Emergence, Detection, and Response* (Washington, DC: National Academies Press, 2003), 59–157. The full list of shared risks is: microbial adaptation and change, human susceptibility to infection, climate and weather, changing ecosystems, human demographics and behavior, economic development and land use, international travel and commerce, technology and industry, breakdown of public health measures, poverty and social inequality, war and famine, lack of political will, and intent to harm (bioterrorism).
26. Paul Farmer, *Pathologies of Power: Health, Human Rights, and the New War on the Poor* (Berkeley: University of California Press, 2003). See also Amartya Sen, *Poverty and Famines* (Oxford: Clarendon Press, 1981) and *Inequality Reexamined* (New York: Russell Sage Foundation, 1992).
27. See Barry R. Bloom, "Bioterrorism and the University," *Harvard Magazine* 106, no. 2 (2001): 48–52. Bloom balances public health realities against the potential bioterrorist threat.
28. The American Association for the Advancement of Science publishes yearly summaries of federal research and development funding
29. Bernadette Tansey and Erin McCormick, "Feinstein Backs Bill for Government Tracking System," *San Francisco Chronicle*, 12 November 2001.
30. NSDD-189: "Where the national security requires control, the mechanism for control of information generated during federally funded fundamental research in science, technology, and engineering . . . is classification."

31. R. J. Jackson, A. J. Ramsay, C. D. Christensen, S. Beaton, et al., "Expression of Mouse Interleukin-4 by a Recombinant Ectromelia Virus Suppresses Cytolytic Lymphocyte Responses and Overcomes Genetic Resistance to Mousepox," *Journal of Virology* 75 (2001): 1205–1210; J. Cello, A. V. Paul, and E. Wimmer, "Chemical Synthesis of Poliovirus cDNA: Generation of Infectious Virus in the Absence of Natural Template," *Science* 297 (2002): 1016–1018; A. M. Rosengrad, Y. Liu, Z. Nie, and R. Jimenez, "Variola Virus Immune Evasion Design: Expression of a Highly Efficient Inhibitor of Human Complement," *Proceedings of the National Academy of Science* 99 (2002): 8808–8813.
32. Journal Editors and Authors Group, "Statement on Scientific Publication and Security," *Science* 299, no. 5610 (2003): 1149.
33. National Research Council (NRC) Committee on Research Standards and Practices to Prevent the Destructive Application of Biotechnology, *Biotechnology in an Age of Terrorism: Confronting the Dual Use Dilemma* (Washington, DC: National Academy Press, 2003), 14.
34. NRC, *Biotechnology Research in an Age of Terrorism.*
35. The other experiments are those that would confer resistance to therapeutic interventions, enhance pathogen virulence or increase its transmissibility, alter its host range, assist in its capacity to go undetected, or improve its suitability as a weapon agent. NRC, *Biotechnology Research in an Age of Terrorism*, 4.
36. Ibid, 9.
37. Maxine Singer, "The Challenge to Science: How to Mobilize American Ingenuity," in Strobe Talbott and Nayan Chanda, eds., *The Age of Terror: America and the World After September 11* (New York: Basic Books, 2001), 195–218.
38. NRC, *Biotechnology Research in an Age of Terrorism*, 14.
39. Daniel Patrick Moynihan, *Secrecy: The American Experience* (New Haven, CT: Yale University Press, 1998), 227.

INDEX